From SHIVA to SCHRÖDINGER

From SHIVA to SCHRÖDINGER

Unravelling Cosmic Secrets with
Trika Shaivism & Quantum Insights

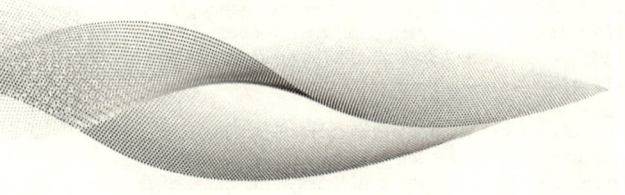

Dr Mrittunjoy Guha Majumdar

HAY HOUSE INDIA
New Delhi • London • Sydney
Carlsbad, California • New York City

Hay House Publishers (India) Pvt Ltd
Muskaan Complex, Plot No. 3, B-2, Vasant Kunj, New Delhi – 110070, India

Hay House LLC, P.O. Box 5100, Carlsbad, CA 92018-5100, USA
Hay House UK Ltd, 1st Floor, Crawford Corner, 91-93 Baker Street, London W1U 6QQ, UK
Hay House Australia Publishing Pty Ltd, 18/36 Ralph St., Alexandria NSW 2015, Australia

Email: contact@hayhouse.co.in
Website: www.hayhouse.co.in

Copyright © Dr Mrittunjoy Guha Majumdar 2024

The views and opinions expressed in this book are the author's own and the facts are as reported by him. They have been verified to the extent possible, and the publishers are not in any way liable for the same.

All rights reserved. No part of this publication may be reproduced, by any mechanical, photographic, or electronic process, or in the form of a phonographic recording, nor may it be stored in a retrieval system, transmitted, or otherwise be copied for public or private use – other than for 'fair use' as brief quotations embodied in articles and reviews – without prior written permission of the publisher.

The author of this book does not dispense medical advice or prescribe the use of any technique as a form of treatment for physical, emotional, or medical problems without the advice of a physician, either directly or indirectly. The intent of the author is only to offer information of a general nature to help you in your quest for emotional, physical, and spiritual well-being. In the event you use any of the information in this book for yourself, the author and the publisher assume no responsibility for your actions.

First published by Hay House India, 2024
10 9 8 7 6 5 4 3 2 1

ISBN 978-93-6611-369-2
ISBN 978-93-6611-732-4 (Ebook)

To Ma and Baba, who inspired me to live up to my name, Mṛtyuñjaya, and to stand firm in my values and principles, even in the face of challenges and setbacks.

To my entangled other, Devyani, who colours my life with tints of love and Śaktī.

May the cosmic dance of Shiva continue to guide us as we seek to understand the essence of being.

Contents

Preface   9

Section I

Chapter 1
The Dance of Cosmic Unity   13

Chapter 2
Doctrine of Dichotomy   29

Chapter 3
Tapestry of Trika Transcendentalism   43

Chapter 4
Convergence of Subject and Object   55

Chapter 5
Echoes of the Empyrean in Pratyabhijñā   70

Section II

Chapter 6
The Quantum Leap Forward   83

Chapter 7
Spooky Action at a Distance   99

Chapter 8
Fleeting Quanta and Fading Waves   118

Chapter 9
Quantum Erasers and Decoherence   132

Chapter 10
The Plinth of Entangled Realities   140

Section III

Chapter 11
Quantrika: Beyond the Veil   149

Chapter 12
Non-Duality and Complementarity   157

Chapter 13
The Illusion of Separation   170

Chapter 14
The Observer's Role in Manifesting Reality   179

Chapter 15
The Nature of Reality   187

Chapter 16
Resonant Portraiture of Episteme   204

Chapter 17
Unveiling Consciousness   217

Acknowledgements   229

References   233

Preface

The advent of quantum physics was radically novel in myriad ways. Physics was no longer deterministic. Reality no longer seemed to be sacrosanct *sans* the observer. It brought with it fundamental questions on the boundaries of a system: how the system interacts with its environment and how agency is exercised in quantum phenomenology. Some of these findings had fundamental resonances with the metaphysics within *Bharatiya Jnana Pranali* (Indian knowledge systems). *Pratyabhijñā*, meaning 'the recognition of the truth of existence,' has been the foundation and focal point of Indic thought. Exploring this fundamental concept of Trika Shaivism is a journey to seek and uncover the ontological foundations of absoluteness, reflexivity, and non-dualism in the universe. The concept of subject-object binary has been of vital importance to the construction of the ideas within Trika Shaivism, as seen in the writings of Utpaladeva (such as in *Iśvara Pratyabhijñā Kārikā*) and Somananda (such as in *Śivadrṛsti*). The binary is ever-so-relevant today, especially with fundamental discussions being undertaken in the field of contemporary science, particularly in the field of quantum mechanics where the question of system-environment interactions and boundaries is being looked at in great detail. The preeminent finding emerging from this meditation is the primacy of correlations in nature, such as with quantum entanglement in the realm of the minuscule. Do the correlations make ours a participatory universe and is there even a way to sharply demarcate a subject-object binary, when it comes to empiricism? This book will explore the intriguing parallels between the metaphysical concepts of Trika Shaivism and the revolutionary principles of quantum mechanics, seeking to uncover the profound connections between these two realms of understanding.

Section I

Section 1

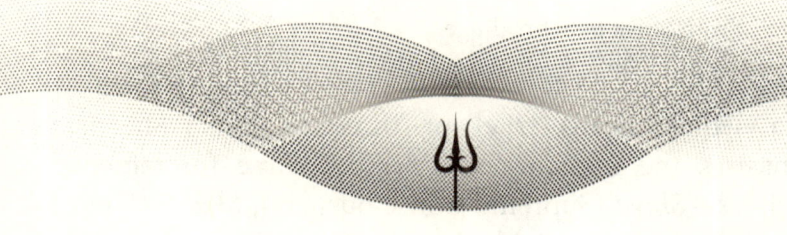

Chapter 1
The Dance of Cosmic Unity

The concept of reality has fascinated humankind since time immemorial. While we can attempt to describe the nature of reality, it remains elusive and beyond full comprehension. In Bharat (India), various philosophical traditions have explored reality in diverse ways: from the non-dualism of Advaita Vedanta, which views reality as a singular, undivided whole, to the categorisation of reality found in Nyāya-Vaiśeṣika philosophy. While non-dualism today is often discussed using via-negativa logic—describing what reality is not—Trika Shaivism, a celebrated school of thought, attributes the essence of reality to dynamism, identifying it, at its most fundamental level, with Shiva. Let's explore the origins and scope of Trika Shaivism and how it expands on the concept of reality and our relationship to it.

Trika Shaivism, also known as Kashmir Shaivism, is a profound and multifaceted philosophical and spiritual tradition that emerged in the Kashmir region of ancient India. With its origins dating back to the early centuries of the Common Era, this tradition has evolved into a rich and diverse school of thought, encompassing a wide range of teachings, practices, and rituals. Derived from the word

'Trika,' which means 'triad' or 'threefold', Trika Shaivism emphasises the interconnectedness of three fundamental principles—*Shiva* (Supreme Consciousness), *Shakti* (Divine Energy), and *anu* (individual soul). These three aspects are considered essential for understanding the nature of reality and the path to spiritual liberation. Shri B. N. Pandit's words are illustrative, albeit somewhat divergent, in this respect:

> 'It is important to understand that, according to Kashmir Shaivism, the analysis of all phenomena into thirty-six *tattvas* (elements of reality) is not an absolute truth. Many practitioners of the Trika system use three tattvas in the process of sadhana: Shiva representing the absolute unity, Shakti representing the link between duality and unity, and Nara representing the extreme duality.'

At its core, Trika Shaivism asserts that the ultimate reality, known as *Paramashiva* or Absolute Consciousness, is the divine essence that permeates everything in the universe. This supreme consciousness is considered transcendent yet immanent, existing both beyond and within the phenomenal world. According to it, the recognition and realisation of this divine consciousness is the key to attaining self-realisation and liberation from the cycle of birth and death. This is encapsulated in the verse that describes Shiva as the embodiment of pure consciousness and bliss.

चिदानन्द रूपः शिवोऽहम्।
शिवोऽहम् चिदानन्द रूपः॥

The philosophy of Trika Shaivism draws heavily from the ancient texts and scriptures of the Shaiva Agamas, which are believed to be the revealed scriptures of Lord

Shiva himself. These sacred texts contain a vast corpus of knowledge, including philosophical treatises, hymns, rituals, and meditative practices. They serve as the foundation for the understanding and practice of this tradition.

One of the central concepts in Trika Shaivism is the idea of *spanda*, which refers to the primordial vibration or pulsation of consciousness. It represents the dynamic interplay of expansion and contraction, creation and dissolution, and is the underlying force behind all manifestations.

Spanda is seen as the dynamic and creative force that gives rise to the entire universe, manifesting as the diverse phenomena we experience and transcending the rational confines of objective reality. It has been beautifully described in the following verse, which speaks of the divine vibration that pulsates in the universe as the conscious essence and also emphasises that understanding and attuning to its presence can help one merge with the state of the eternal Self:

यस्य जगति स्पन्दति चेतनामूर्तिः
सा देवतानां विभुरुच्यते स्पन्दः।
तं ज्ञात्वा स्पन्दितुमनाहविद्यमानः
ब्रह्मात्मभावं प्रविलाप्य सर्वदा॥

In Trika Shaivism, the individual soul, known as anu, is regarded as a microcosmic reflection of the divine consciousness. Each individual is believed to possess the innate potential to realise their true nature as divine beings. However, this realisation is often obstructed by the limitations of the ego, ignorance, and attachment to worldly experiences. The path of Trika Shaivism involves the systematic unravelling of these limitations and the awakening of the dormant divine consciousness within.

Trika Shaivism offers a comprehensive framework for spiritual practice and self-realisation. It encompasses various methods, including mantra recitation, meditation, breath control, ritual worship, and contemplative inquiry. These practices are designed to purify the mind, cultivate heightened awareness, and ultimately lead to the direct experience of the divine. One of the distinguishing features of Trika Shaivism is its emphasis on the direct transmission of spiritual knowledge from teacher to student, or more broadly, experiential accessibility to insights into the nature of reality. This lineage-based transmission, known as *śaktipāt*, involves the transfer of spiritual energy and awakening through the grace of an enlightened master. *Guru-shishya parampara* (the tradition of passing spiritual knowledge from guru to disciple) is considered vital in Trika Shaivism, as the guidance and blessings of a realised teacher are seen as essential for the spiritual aspirant's progress.

In Kashmir Shaivism, the descent of divine grace, or śaktipāt, is categorised into different levels as shown below:

1. **Tīvra-tīvra-śaktipāt:** This highest level results in immediate liberation and oneness with Shiva.

2. **Tīvra-madhya-śaktipāt:** This level leads to self-realised enlightenment without the need for a guru.

3. **Tīvra-manda-śaktipāt:** It creates a strong desire to find a guru, whose mere presence can prompt enlightenment.

4. **Madhya-tīvra-śaktipāt:** This involves seeking a guru's guidance, leading to eventual enlightenment and union with Shiva, often after death.

5. **Madhya-madhya-śaktipāt and Madhya-manda-śaktipāt:** At these lower intensities, the aspirant's desires for worldly pleasures delay liberation, requiring additional reincarnations.

6. **Manda-śaktipāt:** This level represents the lowest intensity, implying slow spiritual progress with liberation occurring over many lifetimes.

Over the centuries, Trika Shaivism has flourished and exerted a significant influence on the cultural, intellectual, and spiritual landscape of Kashmir as well as other regions. It has produced numerous great sages, scholars, and poets who have contributed immensely to the development and articulation of its teachings. The poetry of mystic saints such as Lal Ded, Nund Rishi, and Abhinavagupta has become an integral part of the spiritual and literary heritage of the Kashmiri people. It stands as a profound spiritual tradition that offers a unique and comprehensive approach to self-realisation and the understanding of the nature of reality. Its teachings emphasise the interconnectedness of consciousness, the transformative power of divine energy, and the potential for every individual to realise their innate divinity. With its rich philosophical doctrines, transformative practices, and profound insights into the nature of existence, Trika Shaivism continues to inspire seekers on the path of spiritual awakening and liberation.

Roots of Trika Shaivism: A Journey Through Time and Tradition

From the mid-9th century, the Tantric Śaiva traditions of the Mantramarga evolved from scriptural anonymity into a rich body of Kashmiri exegesis. This literature primarily reveals two schools: the Trika and Krama schools on the left, and the more conservative, Veda-aligned Śaiva Siddhanta on the right.

The Trika and Krama doctrines were definitively formulated by Abhinavagupta (circa 975-1025 CE) and his student Kṣemarāja, while the Śaiva Siddhanta reached its peak with the works of their contemporary, Rāmakaha. Abhinavagupta's tradition, which drew on earlier figures such as Vasugupta (circa 875-925 CE) and Somananda (circa 900-950 CE), presented a counter-movement to the established Śaiva Siddhanta.

Both traditions addressed seekers of liberation (*mumukṣus*) but with differing approaches. The Trika authorities critiqued the Śaiva Siddhanta for its ritualistic emphasis, arguing that it exaggerated the role of rituals in attaining salvation. In contrast, Trika philosophy, as formulated by Abhinavagupta and Kṣemarāja, proposed that liberation could be achieved not only through ritual but also through mystical experience and gnosis. While Śaiva Siddhanta emphasised liberation through ritual purity and a distinct separation from Śiva, Trika focused on an internal realisation of one's divine identity through esoteric practices and meditation. This tradition has attracted numerous scholars, mystics, and spiritual seekers throughout history.

As discussed earlier, Shaivism has a rich history in Kashmir, dating back to ancient times. It is believed that the roots of Shaivism in the region can be traced back to the pre-Vedic era. The earliest evidence of Shaivism in Kashmir can be found in the Naga cult and worship of Shiva as a serpent deity. Additionally, historical records by Kalhana mention that around 371 BCE, King Gopaditya constructed a temple on a hilltop dedicated to Jyesthesvara, also known as Shiva Jyestharuda. Over time, the tradition evolved and assimilated various influences, including Vedic, Buddhist, and tantric philosophies. One of the key figures in the development of Trika Shaivism was Vasugupta, an enlightened sage who lived in the 9th century CE. Vasugupta is credited with codifying the teachings of Shaivism into a systematic philosophical framework known as the Shiva Sutras. These aphorisms, divided into three sections (*pratyabhijñā, vimarsha,* and *prakasha*), form the cornerstone of Trika Shaivism.

Swami Lakshman Joo, mystic and scholar of Kashmir Shaivism, encapsulated the reason for appending 'Kashmir' when referring to Trikadarśana:

> 'Saṅgamāditya came from Kailash Parbat to Kashmir. His lineage continued in Kashmir. Also, Atrigupta came to Kashmir and was accommodated here during the time of Rāja Lalitāditya. His expertise was also brought into the religious landscape of Shaivism in Kashmir. Abhinavagupta was in his lineage as well.'

The aesthetics-inspired experiential spirituality of seers in Kashmir Valley played a major role in helping develop Trika Shaivism. It is said that enchanting verses that comprise the work known as *Shivastotravali* were born on Mahasarit (today known as the Dal Lake) as an inspirational outpouring of devotion of Utpaladeva. The golden era of Trika Shaivism came with the emergence of Abhinavagupta, a polymath, philosopher, and mystic who lived during the 10th and 11th centuries CE. Abhinavagupta expanded upon the foundations laid by Vasugupta and enriched the tradition with his profound insights. He authored numerous works, including the *Tantraloka*, a monumental treatise that elucidates the entire gamut of Trika Shaivism.

Trika Shaivism is renowned for its profound philosophical tenets, which emphasise the ultimate reality of the Supreme Consciousness (*Paramashiva*) and the inherent divinity present in all beings. The tradition posits a non-dualistic worldview, asserting that everything is an expression of Shiva's infinite consciousness. It also places great importance on the concepts of self-recognition (pratyabhijñā) and divine grace (*anugraha*). In addition to its philosophical underpinnings, Trika Shaivism incorporates various ritualistic practices and tantric techniques. These practices involve the use of mantra, yantra, mudra, and meditation to connect with the divine and attain spiritual liberation. It strongly emphasises the direct experience of transcendence and the integration of spirituality into everyday life.

Throughout its history, this tradition faced both external and internal challenges. The external challenges came in the form of Islamic invasions in the 14th century, which significantly disrupted the spiritual fabric of the region. However, despite these challenges, the teachings of Trika Shaivism were preserved by dedicated practitioners and later experienced a revival in the 20th century.

Unveiling a Dynamic Oneness: Core Beliefs of Trika Shaivism

Trika Shaivism acknowledges the multifaceted nature of reality but ultimately asserts the underlying non-dualistic nature of all existence. This philosophy aligns with the *Advaita Siddhanta* (non-dualistic) philosophy, which posits that the apparent duality in the world is an illusion. It suggests that there is only one ultimate reality, referred to as *Paramashiva*, and everything else is a manifestation of this Supreme Consciousness. This non-dualistic perspective transcends conventional boundaries and unifies the divine and the human. In the *Nirvāṇaṣatkam* by Sri Adi Shankaracharya, the following verse speaks of the ultimate reality as neither the mind, intellect, ego, nor the reflections of the inner self, nor the five senses.

मनोबुद्ध्यहङ्कार चित्तानि नाहं न च श्रोत्रजिह्वे न च घ्राणनेत्रे।
न च व्योम भूमिर्न तेजो न वायुः चिदानन्दरूपः शिवोऽहम् शिवोऽहम्॥

This means that the ultimate reality is neither ether, earth, fire, nor wind, but something beyond these elements. It is often regarded as eternal knowing, being-ness, and bliss—an experience that defies articulation.

According to Trika Shaivism, consciousness is the fundamental essence of all existence. It asserts that Supreme Consciousness is omnipresent, manifesting in various forms, and emphasises the importance of self-recognition and realising one's true nature as the Supreme Consciousness. Put simply, human beings, as individual souls, are considered microcosmic reflections of the macrocosmic consciousness.

The following verse from *Spandakārikā* (*Svarūpaspanda* 24) expands on this concept, particularly in terms of vital energies. It discusses how by taking refuge in the supreme state of spanda, both *apāna* and *prāṇa* (vital energies) converge within *suṣumnā*, eventually rising upward and abandoning the empirical universe (*brahmāṇḍa*) to dissolve into the ultimate reality.

तामाश्रित्योर्ध्वमार्गेण चन्द्रसूर्यावुभावपि।
सौषुम्नेऽध्वन्यस्तमितो हित्वा ब्रह्माण्डगोचरम्॥

Trika Shaivism celebrates this realisation through creative expressions of the divine in art, music, dance, and poetry.

The aesthetic manifestation of *anuttara* provides a powerful medium for connecting with the divine and experiencing unity. It encourages individuals to explore their creative potential as a means of spiritual growth and self-expression.

Trika Shaivism also recognises the vital role of Shakti, the divine feminine energy or power that animates and sustains all existence. It is considered the active aspect of consciousness, while Shiva represents pure, transcendent awareness. The following verse from *Spandakārikā* (*Svarūpaspanda* 16) highlights the power of Shakti, residing within and binding the conditioned being.

सेयं क्रियात्मिका शक्तिः शिवस्य पशुवर्तिनी।
बन्धयित्री स्वमार्गस्था ज्ञाता सिद्ध्युपपादिका॥

Recognising the aspect of Shakti enables us to move towards our own Self. This acknowledgement of the feminine principle highlights the importance of balance and harmony between the masculine and feminine energies within individuals and society.

In conclusion, Trika Shaivism provides a holistic approach to spiritual evolution, emphasising self-recognition, the interplay of consciousness and energy, and the cultivation of creativity. Its non-dualistic worldview offers a profound understanding of the nature of reality and the individual's place within it. By recognising the inherent unity of all beings, this philosophy transcends conventional dualities and invites individuals to embrace their divine nature.

Trika Shaivism in Context: Mapping the Spiritual Landscape

Shaivism, one of the major schools of Hinduism, encompasses a diverse range of philosophical and spiritual traditions, such as Shaiva Siddhanta, Kashmir Shaivism, Pashupata, Lingayat, Nathpanthi, Aghora, and Kapala. Among its various branches, Trika Shaivism stands out for its unique features and distinct perspectives. In this section, we will explore the key principles and distinctive aspects of this tradition by comparing it to other branches of Shaivism. By understanding the similarities and differences, we can gain deeper insights into its profound teachings.

As discussed earlier, Shaivism represents the worship of Lord Shiva, the Supreme Being, and is one of the oldest and most prominent branches of Hinduism. It encompasses a rich diversity of theological, philosophical, and ritualistic practices across different regions and time periods. For instance, Shaiva Siddhanta, prevalent in South India, adopts a dualistic approach, viewing Shiva as the supreme deity distinct from individual souls and emphasising a realistic understanding of the material world. In contrast, Kashmir Shaivism, especially the pratyabhijñā school, embraces a non-dualistic and idealist perspective, asserting that the individual self is ultimately identical to Shiva, who is considered the sole reality.

On the other hand, the Pāśupata school, one of the oldest Shaiva traditions originating in the 2nd century CE, focuses on ascetic practices and the worship of Shiva as the lord of all creatures. It promotes a dualistic understanding of reality, distinguishing between the individual soul and the Supreme Being. Moreover, the Pāśupata tradition comprises two principal branches: the Mahāpāśupata and the Lakula Pāśupata, associated with Lakulisa. Lakulisa's doctrine, known as Ishvara Kartri Vadaha (the creative power of the sovereign being), is detailed in the *Gana Karika* and its commentary, *Panchartha Bhashya*. Lakulisa reinterpreted core concepts of the Maheshwara doctrine—such as *Pati* (Supreme God), *pashu* (individual soul), *pasha* (bonds), *karana* (cause), and *karya* (effect)—and emphasised distinct practices for each stage of spiritual development.

The Lingayat tradition, founded by the 12th-century philosopher Basava, centres on the worship of the *linga* and advocates for an egalitarian approach to spirituality, rejecting the caste system. It emphasises devotion to Lord Shiva through the formless symbol of the linga and upholds the concept of qualified non-dualism, recognising the eternal relationship between the individual soul and Shiva. Lingayatism outlines a spiritual path called the Satsthalasiddhanta, which comprises six progressive stages. The journey begins with the individual as a devotee, advances to becoming a master and a receiver of divine grace and continues with recognising the divine presence in one's life breath. This path leads to a phase of surrender, where one experiences the oneness of God and the Self. Ultimately, it culminates in the complete union of the soul with God, achieving liberation (*moksha*). This progression reflects a deepening of devotion (*bhakti*), evolving from external worship to an abstract awareness of the divine, and finally to the realisation of oneness with God.

Additionally, the Nathpanthi sect, which emerged in the 8th century CE, integrates elements of yoga and mysticism, focusing on personal experience and direct communion with the divine. In contrast, Aghora represents a more radical and esoteric approach, often involving unconventional practices designed to transcend societal norms and achieve liberation.

As we can see, each of these schools, while centred on the worship of Shiva, presents unique interpretations and methodologies for understanding the divine and attaining spiritual liberation. Trika Shaivism occupies a distinct

Trident Yantra of Parama Siva, Image by Visarga, licensed under CC BY-SA 3.0

position within the Shaivite traditions. Its primary emphasis is on the non-dualistic nature of reality and the individual's relationship with the Supreme Consciousness. This non-dualistic perspective differentiates Trika Shaivism from other Shaivite branches, which may focus on dualism or qualified non-dualism.

Trika Shaivism stands out among the branches of Shaivism due to its non-dualistic worldview and emphasis on the unity of the individual soul with the Supreme Consciousness. While other branches may lean towards dualism or qualified non-dualism, Trika Shaivism challenges these distinctions and offers a profound understanding of the nature of reality and spiritual liberation. Its unique features, such as the concept of spanda, recognition of divine energy, the path of pratyabhijñā, and the celebration of aesthetic expression, contribute to its distinctiveness within the Shaivite traditions. By exploring and appreciating these unique aspects, one can delve deeper into the transformative teachings of Trika Shaivism, which lead to self-realisation, as put forth in the *Śivasūtra* (22), which states that by uniting with the 'Great Lake' (Shiva), the yogī can have the experience of the potency of all the mantras.

महाह्रदानुसन्धानान्मन्त्रवीर्यानुभवः

The most intriguing Shloka in the *Śivasūtra* is (36), which suggests that when difference or duality is suppressed or rejected, an action or agency exercised becomes an act of creation.

भेदतिरस्कारे सर्गान्तरकर्मत्वम्

In Indian knowledge systems, particularly Advaita Vedanta, the notion of non-duality (Advaita) posits that the ultimate reality is undivided and indivisible, where the apparent separation between the individual self (*atman*) and the universal consciousness (*brahman*) is an illusion (*maya*). This perspective aligns with the concept that when the illusion of separation fades away, the underlying unity of existence reveals itself. Modern physics, especially quantum mechanics, echoes this sentiment through phenomena such as quantum entanglement, where particles remain interconnected regardless of spatial separation, suggesting a fundamental oneness at the subatomic level. This interconnectedness challenges the classical notion of isolated systems and implies that actions at one level can have creative consequences throughout the entire system. Thus, both Indian philosophy and modern physics converge on the idea that at a fundamental level, there is no separation—only unity. The key insight here that I would like to posit is as follows: Correlations can facilitate the transition of action to novel phenomena.

The natural world is full of fascinating examples of organisation and emergence, where many individual components come together to form something entirely new and remarkable. It is almost as if the whole is greater than the sum of its parts. Take the human brain, for instance. Each individual neuron is a relatively simple unit, but when millions of them start exchanging information in a highly coordinated way, the result is the incredible ability to think, reason, and experience consciousness. It is mind-boggling to consider that this capacity arises not from a single neuron, but from the collective, organised activity of the brain as a whole. As Nobel Laureate Eric Kandel elucidated:

'The task of neural science is to explain behaviour in terms of the activities of the brain. How does the brain marshall its millions of individual nerve cells to produce behaviour, and how are these cells influenced by the environment . . . ? The last frontier of the biological sciences—their ultimate challenge—is to understand the biological basis of consciousness and the mental processes by which we perceive, act, learn, and remember.'

A similar principle plays out in certain advanced materials that we have engineered. Electrons, those tiny subatomic particles, exhibit intricate 'dancing' patterns with each other, avoiding the strong repulsive forces that would normally keep them apart. This electron choreography can give rise to astounding emergent properties, like superconductivity which is the ability for electricity to flow with zero resistance. The discovery of superconductivity in iron-based materials back in 2008 was a real game-changer. Despite iron's typical magnetic properties, which would typically interfere with superconductivity, these materials have somehow achieved it. This has left scientists scratching their heads, but also brimming with excitement—could this breakthrough unlock even higher temperature superconductors that could revolutionise technology? Researchers are now delving into unravelling the structural secrets and electronic behaviours of these novel superconductors. And they are not stopping there; they are also exploring other exotic states of matter, like topological insulators, where electrons can flow freely on the surface but become trapped inside. It is like an electronic version of a Möbius strip. To peer into these quantum-scale mysteries, scientists are pulling out all the stops by using powerful magnetic fields, temperatures near absolute zero, and even extreme pressures to coax these materials into revealing their most closely guarded secrets. Who knows what other wonders of organisation and emergence are waiting to be discovered in the vast, uncharted territory of the quantum world?

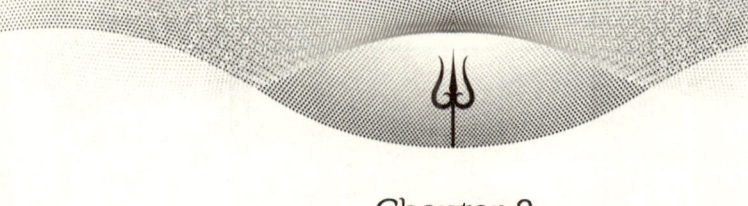

Chapter 2
Doctrine of Dichotomy

General philosophy encompasses a wide range of fundamental questions about the nature of reality, knowledge, and existence. One concept that has persisted throughout philosophical discourse is the subject-object binary. This binary refers to the perceived separation between the subject (the observer, thinker, or perceiver) and the object (the observed, thought about, or perceived). The subject-object binary plays a crucial role in shaping our understanding of reality, consciousness, and our place in the world. There is a common misconception that the subject-object binary is deeply rooted in Western philosophical traditions as it has been explored by philosophers such as René Descartes, Immanuel Kant, and Edmund Husserl. At its core, this binary suggests a distinction between the perceiving subject and the perceived object. It asserts that the subject exists independently of the object and that knowledge of the object is acquired through the subject's sensory experiences.

One of the most prominent philosophical frameworks associated with the subject-object binary is dualism. Dualism posits a separation between mind and matter, asserting that consciousness or the mind is distinct from

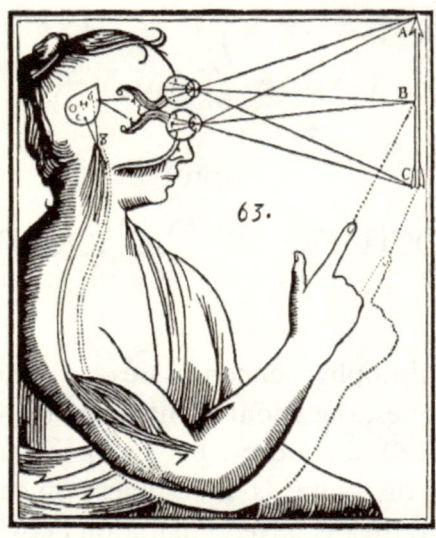

Descartes Mind and Body, Image by René Descartes (public domain)

the physical body. The concept of dualism in philosophy has been a subject of extensive debate, with influential thinkers offering diverse perspectives on the mind-body relationship. Aristotle, building upon Plato's views, proposed a hierarchical arrangement of souls, with the nutritive, perceptive, and rational faculties corresponding to plants, animals, and humans respectively. He believed the soul is the hylomorphic form of a viable organism, where each level formally supervenes upon the preceding one. For Aristotle, the first two souls, rooted in the body, perish with the organism, while the intellective part of the mind remains immortal. In contrast, Plato advocated for the soul's independence from the physical body, espousing the concept of metempsychosis, the migration of the soul to a new body. In fact, he went so far as to talk of a tripartite soul, as highlighted in the book, *Plato and the Divided Self:*

'Plato's account of the tripartite soul is a memorable feature of dialogues like the Republic, Phaedrus, and Timaeus. It is one of his most famous and influential yet least understood theories. It presents human nature as both essentially multiple and diverse, and yet somehow also one, divided into a fully human "rational" part, a lion-like "spirited" part and an "appetitive" part likened to a many-headed beast. How these parts interact, how exactly each shapes our agency, and how they are affected by phenomena like eros and education is complicated and controversial.'

This view has been criticised by some as a form of reductionism, potentially overlooking important variables. Dualism is often juxtaposed with monism, which posits a single fundamental category of existence. Substance dualism contrasts with materialism, while property dualism can be seen as a form of emergent materialism or non-reductive physicalism. René Descartes, a prominent dualist, famously argued for a division between the thinking subject and the material world. Descartes' formulation of the subject-object binary laid the groundwork for centuries of philosophical inquiry.

However, dualism has faced criticism for its inability to reconcile the subjective experiences of consciousness with the objective nature of the physical world. Philosophers like Gilbert Ryle and Thomas Nagel have argued that consciousness is not a separate substance but rather an integral aspect of our existence, challenging the strict subject-object binary. Additionally, phenomenology, developed by Edmund Husserl, offers an alternative perspective on the subject-object binary. Phenomenology seeks to describe and analyse subjective experience as it is directly given. Husserl argued for the reduction of external influences to focus on the pure essence of consciousness, suspending the assumptions of the subject-object binary. According to phenomenology, the subject and object are inherently interconnected. The object does not exist independently of the subject; rather, it is constituted through the intentional acts of consciousness. In other words, the subject and object co-constitute each other in a mutually dependent relationship. This perspective challenges the dualistic understanding of the subject-object binary and provides a more nuanced understanding of consciousness and experience.

While the subject-object binary has been a fundamental framework in philosophy, it has faced significant critiques. Poststructuralist thinkers such as Michel Foucault, Jacques Derrida, and Judith Butler have questioned the binary's claim to universality and inherent stability. They argue that the binary reinforces power dynamics, marginalises certain perspectives, and limits our understanding of reality. In fact, Foucault has a fascinating definition of the individual, somewhat aligned with what is outlined in *Śivasūtra*.

> '. . . it's my hypothesis that the individual is not a pre-given entity which is seized on by the exercise of power. The individual, with his identity and characteristics, is the product of a relation of power exercised over bodies, multiplicities, movements, desires, forces.'

Foucault, in his analysis of power relations, suggests that the subject-object binary is shaped by social and historical contexts. He argues that power relations determine how subjects are positioned and how objects are constructed. Similarly, Derrida deconstructs the binary by highlighting the inherent instability of language and meaning. He contends that language itself is a play of differences, constantly challenging fixed categories such as subject and object. He famously said, talking of his theory of deconstruction:

> 'That is what deconstruction is made of: not the mixture but the tension between memory, fidelity, the preservation of something that has been given to us, and, at the same time, heterogeneity, something absolutely new, and a break.'

Contemporary philosophical and scientific developments have expanded our understanding of the subject-object binary. For instance, cognitive science and neuroscience provide insights into the interplay between subjective experience and objective reality. The concept of embodied cognition suggests that our bodily interactions with the world shape our understanding of it, challenging the strict separation between subject and object. Furthermore, feminist philosophy and critical race theory have drawn attention to the ways in which the subject-object binary can perpetuate systems of oppression. These perspectives emphasise the importance of acknowledging the situatedness of individuals, their embodied experiences, and the power dynamics that influence knowledge production. Eric Micha'el Leventhal gave a self-looped structure to consciousness, thereby highlighting an important and interesting aspect of consciousness.

> 'The awareness that seeks to know is the very object of its own seeking.'

Notwithstanding the reflexive nature of consciousness, we can see that the subject-object binary has been a prominent feature of general philosophy, influencing our understanding of reality, knowledge, and consciousness. While the binary has faced significant critique, it continues to shape philosophical inquiry and remains a point of contention and exploration. By engaging with alternative perspectives, such as phenomenology, poststructuralism, and emerging interdisciplinary fields, we can move beyond the limitations of the subject-object binary and develop a more comprehensive understanding of our existence in the world. Acknowledging the interconnectedness of subject and object allows us to navigate the complexities of human experience, challenge oppressive systems, and promote a more inclusive and holistic understanding of reality.

Implications of the Subject-Object Distinction

The subject-object distinction is a fundamental concept that has intrigued thinkers for centuries. In this section, we will explore the ontological and epistemological implications of this distinction, delving into its significance for our understanding of reality and knowledge acquisition.

A subject is an entity that possesses agency, experiences consciousness, and is situated in relation to external entities. Conversely, an object is anything that a subject observes or experiences, including other beings. In essence, a subject is the observer, while an object is what is observed. And though this distinction has evolved over time, in the 20th century, philosophers expanded on the traditional subject-object dichotomy with concepts such as John Searle's 'institutional facts'. These concepts explore a middle ground between subjective and objective realities, characterised as 'ontologically subjective, epistemologically objective'. For example, the status of a note as money is not an inherent objective fact but is recognised as such through collective agreement. This illustrates the complex interaction between subjective perceptions and objective realities.

In other words, the ontological perspective of dualism posits a strict separation between the subject and the object. It asserts that there are two fundamentally different substances: the mind (subject) and the physical world (object). This view, advocated by philosophers such as René Descartes, perceives the subject and object as distinct entities with little interaction between them. In contrast, monism challenges the subject-object distinction by proposing a unified nature of reality. Monistic philosophies, such as idealism, argue that everything we perceive as objects is ultimately a creation of the mind.

According to idealists like George Berkeley, the existence of objects depends on their being perceived by a subject. From this standpoint, the subject-object relationship becomes more entangled, with the subject playing a crucial role in the creation and interpretation of the object. The subject-object distinction also has profound implications for our understanding of knowledge acquisition. Epistemology, the branch of philosophy concerned with the nature and limits of knowledge, is closely intertwined with this distinction. Empiricism, an epistemological approach associated with philosophers like John Locke and David Hume, emphasises the role of the senses in acquiring knowledge. Empiricists argue that our knowledge of the world is derived primarily through direct observation and experience of objects. According to this view, the subject passively receives information from the object, assuming a clear distinction between the knower and the known. Contrastingly, rationalism takes a different stance by highlighting the role of reason and innate ideas in the acquisition of knowledge. Rationalists, such as René Descartes and Immanuel Kant, argue that certain truths can be known a priori, independent of sensory experience. For rationalists, the subject is an active participant in the process of knowledge acquisition, shaping and interpreting the objects based on innate mental structures. René Descartes famously said:

> 'I suppose therefore that all things I see are illusions; I believe that nothing has ever existed of everything my lying memory tells me. I think I have no senses. I believe that body, shape, extension, motion, location are functions. What is there then that can be taken as true? Perhaps only this one thing, that nothing at all is certain.'

This highlights the non-absoluteness of reality and the elements therein. In *An Answer to the Question: What is Enlightenment?*, Immanuel Kant said:

> 'Enlightenment is man's release from his self-incurred tutelage. Tutelage is man's inability to make use of his understanding without direction from another. Self-incurred is this tutelage when its cause lies not in lack of reason but in lack of resolution and courage to use it without direction from another. *Sapere aude!* (dare to think) "Have the courage to use your own reason"—that is the motto of enlightenment.'

While the subject-object distinction has shaped philosophical discourse, it is not without its critiques and challenges. Some argue that the distinction itself is artificial and arbitrary, potentially limiting our understanding of reality and knowledge. Postmodernist philosophers, for instance, question the existence of a fixed subject-object relationship, emphasising the role of language and social constructs in shaping our perception of reality. They contend that our understanding of the world is mediated through cultural and linguistic frameworks, challenging the notion of an objective reality independent of the subject. This perspective challenges the enlightenment ideals that aimed to define truth and knowledge as universally and objectively attainable. Postmodernists like Jacques Derrida and Michel Foucault argue that meaning is inherently unstable and shaped by context, historical conditions, and power dynamics. Derrida's concept of 'différance' demonstrates how meaning is perpetually deferred and never fully present, suggesting that our understanding of objects is mediated through language, which is itself open to interpretation and change.

Postmodernists also critique grand narratives that claim to provide absolute truths, arguing that these narratives often suppress diverse perspectives and experiences, leading to a homogenisation of thought. This critique extends to the notion of the 'death of the subject', which views individuals as products of social and linguistic influences rather than as autonomous entities. As a result, postmodern philosophy encourages a re-examination of how knowledge is constructed, advocating for an appreciation of the interplay between subjectivity and objectivity. It emphasises the importance of local, contextual truths over universal claims and promotes a pluralistic approach to knowledge, where multiple interpretations can coexist and reflect the complexities of human experience.

Quantum physics and quantum field theory challenge various traditional ontological and epistemological assumptions by highlighting the interconnectedness of particles. Quantum theory suggests that the observer's presence can influence the behaviour of particles, blurring the lines between the subject and object. In fact, Stephen Hawking qualified this further for any instance besides the current, in his work, *The Grand Design:*

'Quantum physics tells us that no matter how thorough our observation of the present, the (unobserved) past, like the future, is indefinite and exists only as a spectrum of possibilities.'

'The universe, according to quantum physics, has no single past, or history. The fact that the past takes no definite form means that observations you make on a system in the present affect its past.'

In conclusion, the examination of the ontological and epistemological implications of the subject-object distinction reveals the complex and multifaceted nature of these concepts. While some philosophical perspectives emphasise a clear separation between the subject and object, others challenge this distinction by proposing a more entangled relationship. The ontological implications raise questions about the nature of reality, with dualism and monism providing contrasting perspectives on the subject-object relationship. Epistemologically, the subject-object distinction influences our understanding of knowledge acquisition, with empiricism and rationalism highlighting different sources of knowledge. However, the subject-object distinction is not without criticism. Postmodernist perspectives and quantum physics challenge the assumption of a fixed and independent subject and object, highlighting the role of the observer's presence in shaping our understanding of reality. Ultimately, the subject-object distinction continues to spark intellectual inquiry, inspiring philosophers, scientists, and thinkers to delve deeper into the nature of existence and the ways in which we acquire knowledge about the world. By critically examining this distinction, we gain a richer understanding of the complexities inherent in our perceptions and interpretations of reality.

Subject-Object Binary in Indic Philosophical Traditions

It is fascinating to explore perspectives on the subject-object binary, especially through the lens of modern sign theory. Charles Peirce defined semiosis as a triadic interaction involving a sign, its object, and its interpretant. This forms the foundation of his formal semiotics, which he divided into speculative grammar (the elements of semiosis), logical critic (modes of inference), and speculative rhetoric (the theory of inquiry and pragmatism).

In Peirce's framework, a sign represents something and can be interpreted in various ways; it is not limited to symbolic or artificial representations. The object is what the sign refers to, encompassing anything discussable or thinkable. The interpretant is the meaning or implication of the sign, formed through an interpretive process that reflects both the object and previous signs about it.

Moreover, Mario D'Amato's doctrine of signs highlights the infinite chain of interpretation where each sign leads to another. This concept has also been explored by thinkers like Klaus Oehler and Jacques Derrida. D'Amato contrasts this with Buddhist semiotics, particularly from the Mahayana text *Mahayanasutralamkara*, which seeks either perfect realisation or complete cessation of semiosis, aligning with Buddhist soteriology. This exploration of sign theory is one of the many ways Indic knowledge systems, including Trika Shaivism, have addressed both the functional and fundamental aspects of the subject-object binary.

Indic philosophical traditions, including Hinduism, Buddhism, and Jainism, have a rich history of exploring the nature of reality, knowledge, and the self. Central to these traditions is the concept of the subject-object binary, which examines the relationship between the perceiver (subject) and the perceived (object).

In Indic philosophies, the idea of duality, often referred to as *dvaita*, is a common theme, highlighting the apparent distinction between the subject and the object. However, Indic traditions also emphasise the underlying unity or non-duality, known as Advaita, which suggests a profound interconnectedness between the subject and the object.

In Hindu philosophy, the subject-object binary is explored through various schools of thought. Advaita Vedanta, for instance, propounded by philosophers like Adi Shankara, emphasises the ultimate oneness of the self (subject) and the universe (object). According to this perspective, the individual self (*jiva*) and the universal self (*brahman*) are ultimately non-dual, with the perceived distinction arising from ignorance (*avidya*). On the other hand, the Nyaya-Vaisheshika school, with its emphasis on logic and epistemology, focuses on the subject-object relationship as a means to acquire knowledge. This school highlights the role of perception (*pratyaksha*) as a reliable source of knowledge, where the subject perceives the object through the senses.

In Buddhist philosophy, particularly in the Mahayana tradition, the subject-object binary is explored in the context of emptiness (*sunyata*). According to the teachings of Nagarjuna and other Buddhist philosophers, the ultimate nature of reality is empty of inherent existence. This implies that the subject and the object lack independent and fixed identities, challenging the notion of a rigid distinction between them. Buddhist philosophy also emphasises the interdependent nature of all phenomena, known as 'dependent origination' (*pratītyasamutpāda*). From this perspective, the subject and the object arise in dependence on each other, and their boundaries become fluid and contingent upon various causes and conditions. Jain philosophy, rooted in the teachings of Mahavira, approaches the subject-object binary from the perspective of non-absolutism (*anekantavada*). According to Jainism, reality is multifaceted and complex, and no single viewpoint can capture its entirety. The subject and the object are seen as interrelated but with multiple aspects or perspectives that can only be comprehended through a comprehensive understanding of reality. Jain philosophy also emphasises the importance of a detached and non-attached perspective (*vairagya*) towards both the subject and the object. By cultivating a state of equanimity and non-attachment, one can transcend the limitations imposed by the subject-object binary and attain a deeper understanding of reality.

The exploration of the subject-object binary in Indic philosophical traditions has significant implications for our understanding of self, reality, and knowledge. It challenges the conventional notion of a rigid distinction between the subject and the object, offering a more holistic and interconnected perspective. By recognising the interdependence and non-duality between the subject and the object, these traditions promote a sense of unity, compassion, and empathy. They invite individuals to go beyond the limitations of dualistic thinking and cultivate a deeper understanding of the interconnectedness of all beings and phenomena. Furthermore, the subject-object binary in Indic traditions encourages an experiential and introspective approach to knowledge. Practices such as meditation, self-inquiry, and mindfulness play a vital role in transcending the limitations of the dualistic framework and gaining direct insights into the nature of reality. In the Srimad Bhagavad Gita (12.12), it is stated that superior to mechanical practice is knowledge, and superior to knowledge is meditation.

श्रेयो हि ज्ञानमभ्यासाज्ज्ञानाद्ध्यानं विशिष्यते

The subject-object binary in Indic philosophical traditions also offers a profound and nuanced understanding of the relationship between the perceiver and the perceived. While acknowledging the apparent distinction between the subject and the object, these traditions also emphasise the underlying unity, interconnectedness, and non-duality. Hindu, Buddhist, and Jain philosophies provide diverse perspectives on the subject-object binary, exploring concepts such as advaita, sunyata, and anekantavada. These philosophical traditions encourage individuals to transcend dualistic thinking, cultivate compassion and empathy, and develop a more holistic understanding of reality. By embracing the insights from Indic philosophical traditions, we can gain a deeper appreciation of the interplay between the subject and the object, and ultimately, enrich our perception, knowledge, and relationship with the world around us.

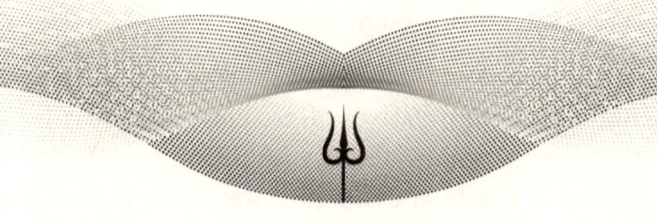

Chapter 3
Tapestry of Trika Transcendentalism

𝒜 key concept in Trika Shaivism is the notion of consciousness. Trika Shaivism places a strong emphasis on the primacy of direct experience and the realisation of the divine essence that permeates all existence. This tradition is rooted in the belief that ultimate reality is not separate from the individual, and the goal is to realise the oneness of the self with the divine. Moreover, in Trika Shaivism, consciousness is regarded as the fundamental essence of all existence. It is the underlying principle that gives rise to the entire universe and is not confined to human beings; rather, it is present in all sentient and insentient entities.

Trika Shaivism asserts that consciousness is not something we possess but rather what we are at the deepest level of our being. It delineates three levels of consciousness: parā, parā-aparā and aparā. These levels represent progressively denser manifestations of consciousness. Parā is the highest level of consciousness in Trika Shaivism. It transcends all dualities and is pure, unmanifest consciousness. It is the supreme reality that underlies everything in the universe. Parā-aparā is the intermediate level of consciousness. It is the bridge between the absolute and the relative. Here, consciousness becomes aware of its own potential for manifestation. Aparā is the lowest level of consciousness. This is the level of manifestation where individual beings, objects, and experiences arise, and it represents the diversity and multiplicity within the universe.

In the *Pratyabhijñāhṛdayam,* Kṣemarāja emphasises the underlying unity not only among all entities in the universe but also across various systems of knowledge. He asserts that the diverse positions of all philosophical systems (*darśanas*) are merely different expressions or roles assumed by the ultimate Self or consciousness.

तद्भूमिकाः सर्वदर्शनस्थितयः

All in all, according to Trika Shaivism, the entire universe is a play of consciousness. The Supreme Consciousness, or Shiva, assumes multiple forms and experiences various states of consciousness to express and explore its own infinite potential. Every individual being is considered a divine expression of consciousness, engaged in the divine play. Trika Shaivism also states that awareness is the key to unlocking the deeper dimensions of consciousness. It is through awareness that one can transcend the limitations of the mind and ego and realise the true nature of self. By cultivating self-awareness, individuals can become conscious participants in the divine play, recognising their inherent divinity. Thereby, meditation holds a crucial place in Trika Shaivism as a means to access and experience higher states of consciousness. Through various meditation techniques, individuals can quiet the mind, dissolve the ego, and merge with the universal consciousness. Simply put, it serves as a direct path to self-realisation and union with the divine. The following extract from *Śivasūtra* (15) describes how by focusing the mind on the essence of consciousness, one can perceive all phenomena—both external and internal—as well as the void (*svāpa*), in their true nature.

हृदये चित्तसङ्घट्टाद्दृश्यस्वापदर्शनम्

The notion of consciousness in Trika Shaivism offers profound insights into the nature of reality and the self. It teaches that consciousness is not separate from the world but is intimately intertwined with it. This realisation of non-dual consciousness is considered the ultimate goal of spiritual practice in Trika Shaivism. By exploring consciousness, practitioners of this tradition can embark on a transformative journey of self-realisation, leading to a deeper understanding of their place in the universe. This understanding of consciousness also has practical implications for daily life. It encourages individuals to see the divinity in themselves and others, fostering compassion, love, and unity. In essence, Trika Shaivism provides a rich philosophical and spiritual framework for seekers to explore the depths of consciousness and discover the ultimate truth of existence.

Subject-Object Binary in the Context of Consciousness within Trika Shaivism

The *Mālinīvijayottaratantra*, considered the most authoritative tantra by Abhinavagupta, is a central text among the 64 non-dual *bhairava-tantras* that form the foundation of the Trika system in Kashmir Shaivism. In his monumental work Tantrāloka, which spans over 6,700 verses, Abhinavagupta highlights the *Mālinīvijayottaratantra* as essential for elucidating Trika teachings. This text synthesises centuries of Shaivite thought into a cohesive philosophical and practical framework. The *Mālinīvijayottaratantra* presents a nuanced philosophical structure that addresses the subject-object binary through its non-dualistic doctrines. Alternatively, Trika Shaivism, tooted in the ancient Shaiva Agamas, offers a comprehensive understanding of consciousness, its manifestation through decentralised elements, and its connection to the external world.

This section delves into the concept of consciousness as elucidated by Trika Shaivism, examining how it sheds light on the intricate relationship between the individual self and universal consciousness.

As mentioned earlier In Trika Shaivism, consciousness is regarded as the fundamental essence of reality, the source and substratum of all existence. It is not confined to the individual mind but pervades the entire universe. Trika Shaiva philosophers emphasise that consciousness is not an object of perception but the subject that perceives; it is the ultimate reality that transcends all dualities and divisions.

Furthermore, the subject-object binary within consciousness refers to the distinction between the perceiver (subject) and the perceived (object). According to Trika Shaivism, this binary is an inherent aspect of consciousness and is responsible for the creation and experience of the phenomenal world. The *Śivasūtra* describes how the awareness of pure consciousness dissolves all conditionings and constraints. In this context, the subject represents pure consciousness or the divine essence, while the object symbolises the external world and its manifestations.

तदारूढप्रमितेस्तत्क्षयाज्जीवसङ्क्षयः

Trika Shaivism teaches that the subject-object binary creates the illusion of duality, leading individuals to perceive themselves as separate from external reality. This perceived separation fosters attachment, desire, and suffering. Trika Shaiva teachings emphasise that recognising the illusory nature of the subject-object binary is essential for transcending it and realising the underlying unity of consciousness. This unity is highlighted at the beginning of the *Ṣaṭtriṁśattattvasandoha*, a significant text by Amṛtānandanātha.

यदयमनुत्तरमूर्तिर्निजेच्छयाखिलमिदं जगत्स्रष्टुम्।
पस्पन्दे स स्पन्दः प्रथमः शिवतत्त्वमुच्यते तज्ज्ञैः॥

The verse explains that *anuttaramūrti*, the highest reality, chose to create the world through its own will. This first vibration, according to those who understand anuttaramūrti, represents the essence of Shiva (*Śiva-tattva*). This non-dualistic perspective, known as Advaita, suggests that the subject and object are ultimately inseparable. Trika Shaiva philosophers assert that realising this inherent unity is essential for transcending dualistic perceptions and achieving enlightenment.

The *Spandakārikā* (1.9) illustrates this concept by stating that when a person's agitation—caused by impurity and desire—completely dissolves, they attain the supreme state of consciousness.

निजाशुद्ध्यासमर्थस्य कर्तव्येष्वभिलाषिणः।
यदा क्षोभः प्रलीयेत तदा स्यात्परमं पदम्॥

This concept has been beautifully captured in the *Śivasūtra*, which describes how the dissolution of perceived differences allows the enlightened seeker of Shiva to realise their inherent creative power.

भेदतिरस्कारे सर्गान्तरकर्मत्वम्

In the context of the subject-object binary, Trika Shaivism also introduces the concept of Shakti, the dynamic creative power of consciousness. Shakti is the force that enables the manifestation of the external world. The central role of Shakti in this framework is vividly illustrated in the following verse from the *Cidgaganacandrika* by Kālidāsa:

अष्टयोनिदलपत्रपूरुषं त्वन्तवृत्तिदलशक्तिपेटिका।
त्वं पतिश्च तव लिङ्गविग्रहो रुद्रसिद्धमुनिवृन्दवन्दिते॥

In this verse, it is stated that the jiva (individual soul) is established through the *mātṛka śaktis* (cosmic energies or divine powers), which are fundamental to the functioning and manifestation of the universe. These energies enable various actions and agencies, which are represented through *bāhya* (external faculties) and *antaḥkaraṇas* (internal faculties). These diverse śaktis are organised within a structured system, with Kālī serving as the encompassing vessel that contains them. In this framework, Śiva is identified as Kameśvara (the lord of desires), Śakti as Kameśvari (the goddess of desires), and their divine union gives rise to the triadic principle of *Nara-Śakti-Śiva* in manifestation. In this perspective, the entire universe, encompassing both immovable and movable entities, is believed to emerge from *akṣara*.

In Indian philosophy, *akṣara* represents the imperishable or eternal, symbolising that which transcends the cycle of birth and death. It is often equated with the ultimate reality or brahman, which is unchanging and beyond physical attributes. The *Bṛhadāraṇyikopaniṣat* describes *akṣara* as formless and eternal, being 'not coarse, not particulated, not short, not long'. In tantric texts, *akṣara* is seen as the foundational principle of creation, the primordial sound or vibration that initiates the creative process. Additionally, akṣara refers to sacred Sanskrit letters or syllables, which are believed to hold specific energies and play a crucial role in the formation of mantras.

The subject-object binary within consciousness has significant practical implications for Trika Shaiva practitioners. By understanding the underlying unity of all things, individuals cultivate a profound sense of interconnectedness and empathy towards all beings. This realisation fosters a balanced approach to life, where one remains engaged with the world while maintaining awareness of the impermanent nature of perceptual objects. Recognising the illusory nature of this binary allows practitioners to transcend duality and experience the unity of consciousness. This entire process of creation, along with the illusory aspects of the universe, is comprehensively addressed in the *Bodhapañcadaśikā*:

अनस्तमितभारूपस्तेजसां तमसामपि।
य एकोऽन्तर्यदन्तश्च तेजांसि च तमांसि च॥

स एव सर्वभूतानां स्वभावः परमेश्वरः।
भावजातं हि तस्यैव शक्तिरीश्वरतामयी॥

शक्तिश्च शक्तिमद्रूपाद्व्यतिरेकं न वाञ्छति।
तादात्म्यमनयोर्नित्यं वह्निदाहिकयोरिव॥

स एव भैरवो देवो जगद्धरणलक्षणः।
स्वात्मादर्शे समग्रं हि यच्छक्त्या प्रतिबिम्बितम्॥

तस्यैवैषा परा देवी स्वरूपामर्शनोत्सुका।
पूर्णत्वं सर्वभावेषु यस्या नाल्पं नचाधिकम्॥

एष देवोऽनया देव्या नित्यं क्रीडारसोत्सुकः।
विचित्रान्सृष्टिसंहारान्विधत्ते युगपद्विभुः॥

अतिदुर्घटकारित्वमस्यानुत्तरमेव यत्।
एतदेव स्वतन्त्रत्वमैश्वर्यं बोधरूपता॥

परिच्छिन्नप्रकासत्वं जडस्य किल लक्षणम्।
जडाद्विलक्षणो बोधो यतो न परिमीयते॥

एवमस्य स्वतन्त्रस्य निजशक्त्युपभेदिनः।
स्वात्मगाः सृष्टिसंहाराः स्वरूपत्वेन संस्थिताः॥

तेषु वैचित्र्यमत्यन्तमूर्ध्वाधस्तिर्यगेव यत्।
भुवनानि तदंशाश्च सुखदुःखमतिश्च या [सुखदुःखमतिर्भवः]॥

यदेतस्यापरिज्ञानं तत्स्वातन्त्र्यं हि वर्णितम्।
स एव खलु संसारो मूढानां यो विभीषकः॥

तत्प्रसादरसादेव गुर्वागमत एव वा।
शास्त्राद्वा परमेशस्य यस्मात्कस्मादुपागतम्॥

यत्तत्त्वस्य परिज्ञानं स मोक्षः परमेशता।
तत्पूर्णत्वं प्रबुद्धानां जीवन्मुक्तिश्च सा स्मृता॥

एतौ बन्धविमोक्षौ च परमेशस्वरूपतः।
न भिद्येते न भेदो हि तत्त्वतः परमेश्वरे॥

इत्थमिच्छाकलाज्ञानशक्तिशूलाम्बुजाश्रितः।
भैरवः सर्वभावानां स्वभावः परिशील्यते॥

सुकुमारमतीञ्शिष्यान्प्रबोधयितुमञ्जसा।
इमेऽभिनवगुप्तेन श्लोकाः पञ्चदशोदिताः॥

In these verses, Bhairava, as described in the Trika Shaiva tradition, represents the ultimate reality that transcends dualities, embodying both creation and destruction. He is the supreme essence from which all beings and the universe arise, sustained by his divine energy (Shakti), which is inseparable from him, much like fire and its burning power. Bhairava's dynamic engagement in the cycles of creation and destruction illustrates his inherent freedom as pure consciousness. The variety of experiences, including pleasure and pain, results from his autonomous actions, while ignorance of this divine nature perpetuates the cycle of worldly existence (*samsara*). Liberation, or *jivanmukti*, is attained through the grace of Bhairava, often facilitated by teachings from a guru or through direct realization of the divine essence. This illustrates that both bondage and liberation are ultimately aspects of the same supreme reality.

Abhinavagupta's teachings also emphasise the importance of meditation and identification with the Supreme. This is further elaborated in the verse from *Tantrāloka* (33.22), which reveals the true nature of the three (trikā) through the internal perception of the Supreme. This process, known as 'unconditioned emanation', describes how the entire universe manifests through a reflection of this supreme reality.

तथान्तःस्थपरामर्शभेदने वस्तुतस्त्रिकम् ।
अनुत्तरेच्छोन्मेषाख्यं यतो विश्वं विमर्शनम् ॥

In conclusion, Trika Shaivism provides a transformative path for practitioners through self-realisation and dedicated practices. This journey fosters a deep understanding of the interconnectedness of all things and leads to a profound shift in perception. By following this spiritual and philosophical framework, individuals can achieve a higher consciousness and uncover the ultimate truth.

The Self and Consciousness in Trika Shaivism

Within Trika Shaivism, the understanding of self and consciousness goes beyond conventional definitions, emphasising the interconnectedness of individual and universal existence. It recognises the self as more than just an individual identity. It views the self as an extension of the divine consciousness, a microcosmic reflection of the universal reality. According to this philosophy, the self is not limited to the physical body or the egoic mind but encompasses the entirety of existence.

Advaita Vedanta identifies fundamental layers of the self, known as the *pancha koshas* or five sheaths. These sheaths represent different levels of experience and awareness. They include the physical body (*annamaya kosha*), the energy body (*pranamaya kosha*), the mental and emotional body (*manomaya kosha*), the wisdom body (*vijnanamaya kosha*), and the bliss body (*anandamaya kosha*). These layers are interconnected, and the realisation of the deeper layers leads to an expanded understanding of the self. Trika Shaivism also points out the illusory nature of the self. It asserts that the conventional understanding of the self as a separate and independent entity is a result of avidya (ignorance) and limited perception. According to its philosophy, the true self transcends these illusory boundaries and is united with the universal consciousness. In *Śivasūtra* (3.7), it is stated that a yogi attains mastery over natural knowledge by completely overcoming maya (ignorance) through a comprehensive victory over hindrances to spiritual realisation.

मोहजयादनन्ताभोगात्सहजविद्याजयः

As mentioned earlier, Trika Shaivism perceives consciousness as the foundational aspect of existence. Consciousness, in Trika Shaivism, is not limited to individual awareness but encompasses the totality of cosmic awareness and pervades all things, from the tiniest atom to the vast expanse of the universe. It delineates three levels of consciousness, known as the states of waking (*jagrat*), dreaming (*swapna*), and deep sleep (*sushupti*). In the broader psychological context, altered states of consciousness—such as hypnagogic (the transition between wakefulness and sleep), hypnopompic (the transition between sleep and wakefulness), and lucid dreaming (awareness during dreaming)—demonstrate the complexity and varied manifestations of consciousness beyond the simple dichotomy of waking and sleeping. Insights from both Trika Shaivism and contemporary psychology underscore the multifaceted nature of consciousness and its profound implications for understanding human experience.

Moreover, these states are not confined to the realm of personal experience but are seen as gateways to understanding the nature of consciousness itself. Trika Shaivism asserts that by transcending these states through meditation and self-inquiry, one can attain the fourth state, *turiya*, which represents pure transcendental consciousness. In the commentary on the *Brihad Bhagavatamrita* by Śrī Śrīmad Bhaktivedānta Nārāyana Gosvāmī Mahārāja, it is mentioned that even Śrī Kṛiṣṇa personally worships Lord Shiva as His intimate devotee (*antaraṅga bhakta*), due to the profound depth of Lord Shiva's realisations and glories.

शिव-दत्त-वरोन्मत्तात् त्रिपुरेश्वरतो मयात्।
तथा वृकासुरादेश् च सङ्कटं परमं गतः॥
शिवः समुद्धृतोऽनेन हर्षितश् च वचोऽमृतैः।
तद्-अन्तरङ्ग-सद्-भक्त्या कृष्णेन वश-वर्तिना।
स्वयम् आराध्यते चास्य माहात्म्य-भर-सिद्धये॥

In conclusion, Trika Shaivism offers profound insights into the nature of consciousness, revealing its intricate and multifaceted aspects. It invites us to transcend the ordinary states of waking, dreaming, and deep sleep to reach the exalted state of turiya, where pure transcendental consciousness resides. This quest for higher awareness is mirrored in the psychological exploration of altered states, illustrating a shared recognition of consciousness's depth and significance. As we delve deeper into these traditions and their modern counterparts, we enhance our understanding of human experience and the vast potential of consciousness.

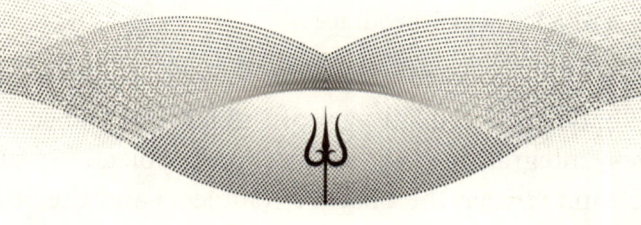

Chapter 4
Convergence of Subject and Object

For a Trika Shaiva practitioner, the manifestation of the universe begins with *parā-vāk* (supreme speech). Śiva, as *parameśvara*, remains transcendent while also acting as the *sākṣī* (witness) and *karta* (agent) who oversees creation. Śiva assumes roles such as *pramātā* (knower) and *prameya* (known) and serves as the means (*karaṇa*) for fulfilling these roles. The process of creation involves the unfolding of thirty-six tattvas (principles), with the first five representing fundamental principles: Śiva, Śakti, Sadāśiva, Īśvara, and Śuddhavidyā. Each of these principles reflects an evolved aspect of Śakti.

In Abhinavagupta's *Paramārthasāra* (6), it is explained that just as a pure crystal reflects different colours, the Lord assumes various forms, such as gods, humans, animals, and trees.

नानाविधवर्णानां रूपं धत्ते यथामलः स्फटिकः।
सुरमानुषपशुपादपरूपत्वं तद्वदीशोऽपि॥

It is important to understand that these tattvas are all aspects of the unified universal subject and object. In

Trika Shaivism, a crucial part of the path to transcendence involves integrating the subject and object, where the distinction between the observer (subject) and the observed (object) is dissolved. This dissolution occurs because Śiva, through his *nigraha śakti* (power of concealment), self-conceals his true nature. This process leads to a direct experience of divine consciousness.

Moreover, Trika Shaivism asserts that the universe is not an illusion, as posited in Advaita Vedanta, but a manifestation of the ultimate reality (*paramashiva*). Trika philosophy describes this manifestation through the concept of *ābhāsa*, where the world is viewed as a real self-projection of Shiva. This differs from the *vivarta* (superimposition) concept in Advaita, which sees the world as an illusion. Trika Shaivism emphasises Shiva's absolute freedom and autonomy to manifest the world, which, while appearing external and different, always exists within Him. In this view, the individual soul (*jiva*) is not separate from brahman but is a manifestation of the same ultimate reality.

In the following sections, we will explore the significance of integrating the subject and object in Trika Shaivism. We will discuss how this integration is achieved and examine the practices that support this union.

From Illusion to Realisation: The Journey of Subject-Object Integration

In conventional understanding, a perceived divide exists between the subject (the observer or experiencer) and the object (the observed or experienced). Trika Shaivism asserts that this division is illusory, stemming from limited perception and that both subject and object arise within the realm of divine consciousness. It also offers practices to integrate the subject and object, helping individuals transcend the illusion of separation to recognise their inherent unity and realise the underlying oneness of all existence. This process involves acknowledging the divine self as the ultimate subject and understanding that all experiences are manifestations of the divine. This recognition enables individuals to shift from a dualistic mindset to one that embraces non-duality and unity.

Let's now explore the key practices in Trika Shaivism for transcending the duality between subject and object.

1. Direct Perception

Trika Shaivism stresses direct perception as the means to integrate subject and object. It encourages practitioners to move beyond conceptualisation and intellectual understanding and instead engage in direct experiential realisation. There is a rather fascinating discussion on the conceptualisation of linga by G. D. Flood, which highlights the importance of perception and insights:

> 'A symbolic form is therefore an expression at one level which discloses a higher level, revealing a structure of reality not immediately apparent. For the Trika, higher realities by which I mean the collective bodies of the universe reveal themselves in symbolic forms and are therefore channels of communication between and within collective bodies. For example, according to Abhinavagupta the term linga—which can designate "symbol" in the sense of an outer emblem displayed by a yogin and the particular "phallic" symbol of Siva—has a manifest and unmanifest or hidden meaning. The symbol is a hierarchical structure whose outer form points to and is derived from its higher, and ultimately its supreme,

form and is thus a channel between layers of the cosmos. Indeed Abhinavagupta classifies the term "symbol" (linga) into the categories of unmanifested (*avyakta*), manifest-unmanifested (*vyaktavyakta*), and manifested (*vyakta*). These form a hierarchical sequence of meaning. The unmanifested symbol (*avyaktalinga*) is equated with the "supreme heart of tranquillity" (*vishrantihrdayam param*) which Jayaratha furthermore equates with other synonyms for the absolute such as awareness of subjectivity (*ahamparamarsha*), the vibration of consciousness (*samvitspanda*), and so on. This is the real meaning of linga for Abhinavagupta leading to true perception to which the manifested or external symbol points and of which it is an expression. Indeed, this unmanifested symbol is defined by Abhinavagupta as that into which this universe is dissolved (*linam*) and which is to be understood as abiding here within (*antahstham*). The manifest-unmanifested symbol is equated with the individual body pervaded by the cosmos (*adhvan*) while the manifested symbol is "a form of vibration which is particularised" (*visheshaspandarupam*), that is, an outer form (*bahirupa*). Thus, the unmanifested symbol corresponds to the supreme body of consciousness, the manifested-unmanifest symbol to the totality of shared realities or collective bodies which make up the universe and are within the individual body, while the manifested symbol corresponds to particular forms external to the individual body.'

This concept of the 'right perception' is central to the thinking of a Trika Shaiva proponent. Through practices such as meditation, contemplation, and self-inquiry, individuals develop the ability to perceive reality directly, free from the filters of the mind.

2. Awareness as Ultimate Subject

According to Trika Shaivism, a key element in the integration of subject and object is awareness. It is through awareness that the subject becomes conscious of the object. However, Trika Shaivism goes beyond this conventional view by asserting that awareness itself is the ultimate subject. By recognising awareness as the core essence of one's being, practitioners dissolve the apparent duality between subject and object and realise the oneness of consciousness.

Vijnana Bhairava Tantra outlines 112 meditation techniques that guide practitioners toward the realisation of non-dual awareness. These methods include yogic breathing, concentration on chakras, and single-point awareness, all designed to help dissolve the boundaries between self and the external world.

3. Tantric Practices

Tantra, by itself, is a highly specialised form of spiritual practice, with elaborate classifications based on modes of spirituality, as highlighted by Alexis Sanderson:

> 'The tantras of Bhairava, so called because they take the form of his answers to the questions of the Goddess (*Devi, Bhairavi*), have been variously listed and classified in different parts of the corpus. Within these tantras there is a primary division between those of the seat of mantras (*mantrapitha*) and those of the seat of *vidyas* (*vidyapitha*). The latter are either union tantras (*yamala-tantras*) or power tantras (*sakti-tantras*). Within the latter one may

distinguish between the tantras of the Trika (or rather of what was later called the Trika) and material dealing with cults of the goddess Kali. Tantras which teach the cult of *tumburu-bhairava* and his four sisters (*jaya, vijaya, Jayanti,* and *aparajita*) are fitted into this scheme as a third division of the vidyapitha... As we ascend through these levels, from the mantrapitha to the yamala-tantras and thence to the Trika and the Kali cult, we find that the feminine rises stage by stage from subordination to complete autonomy."

Trika Shaivism incorporates tantric rituals and symbolism as powerful tools to facilitate subject-object integration. Tantric practices use external objects and rituals to invoke the divine presence within oneself. Through these practices, practitioners can merge their subjective experiences with objective symbols and recognise the divine presence in both.

4. Bhakti

Bhakti, or devotion, is another crucial aspect of Trika Shaivism that fosters the integration of subject and object. By cultivating deep devotion to the divine, practitioners can surrender their individual identity and merge with the object of devotion. This surrender leads to the dissolution of the illusory divide, allowing for a direct experience of the divine as the true subject.

In Trika Shaivism, various forms of bhakti are practised. For example, the Kaula cult focuses on worshipping three goddesses—Para, Parapara, and Apara—arranged in a triangular formation with Kulesvara and Kulesvari at the centre. Worship practices often include external rituals performed on a red cloth, using symbols like vermilion powder and coconuts. The essence of bhakti is beautifully expressed in verse 7 of Puṣpadanta's *Śivamahimnaḥ Stotram*, which describes how different spiritual paths, including the teachings of the three Vedas, *Sāṅkhya*, yoga, the doctrine of *Paśupati* (*Śaiva*), and *Vaiṣṇava* teachings, all lead to the realisation of the Supreme Self or Shiva.

त्रयी साङ्ख्यं योगः पशुपतिमतं वैष्णवमिति प्रभिन्ने प्रस्थाने परमिदमदः पथ्यमिति च।
रुचीनां वैचित्र्यादृजुकुटिलनानापथजुषां नृणामेको गम्यस्त्वमसि पयसामर्णव इव॥

This sense of devotion and surrender is also expressed in Sri Adi Shankaracharya's *Śivamānasapūjā* (verse 1).

रत्नैः कल्पितमासनं हिमजलैः स्नानं च दिव्याम्बरं नानारत्नविभूषितं मृगमदामोदाङ्कितं चन्दनम्।
जातीचम्पकबिल्वपत्ररचितं पुष्पं च धूपं तथा दीपं देव दयानिधे पशुपते हृत्कल्पितं गृह्यताम्॥

In this verse, the spiritual seeker reveres Shiva, the fundamental reality of the universe, by describing an imagined offering. The seeker envisions a seat made of jewels, an ablution with cool water, and divine clothes adorned with various gems. They imagine offering sandalwood scented with musk, a floral arrangement with jasmine, *champaka*, and *bilva* leaves, along with fine incense and a brilliant light. The seeker then asks Shiva to accept these offerings, which are envisioned in their heart, as an expression of devotion to the ultimate reality.

Non-Dual Awareness

Trika Shaivism emphasises the cultivation of non-dual awareness, where the subject-object distinction ceases to exist. Through sustained practice and deepening realizations, individuals shift from perceiving reality through a fragmented lens to experiencing the unbounded unity of consciousness. In this state, subject and object merge into a seamless whole, transcending the limitations of dualistic perception. This concept is emphasised in the *Spandakārikā* (3.19), which states that by being firmly rooted in the principle of spanda, the spiritual seeker gains control over the cycles of disappearance and emergence. As a result, they become the true enjoyer of eternal bliss and attain mastery over spiritual powers (Śakti).

यदा त्वेकत्र संरूढस्तदा तस्य लयोदयौ।
नियच्छन्भोक्तृतामेति ततश्चक्रेश्वरो भवेत्॥

The integration of subject and object is a profound path to transcendence and liberation. By recognising the illusory nature of this divide and engaging in practices that cultivate direct perception, awareness, devotion, and non-dual awareness, practitioners can merge the subject and object into a unified whole. This integration facilitates a direct experience of divine consciousness, resulting in a transformative shift in one's understanding of self and reality.

From Knower to Known: Exploring the Concepts of Pramātā and Prameya

In the rich landscape of Indian philosophy, the concepts of *pramātā* (the subject) and *prameya* (the object) are foundational for understanding perception, knowledge, and reality. In this section, we will delve into the depths of pramātā and prameya, exploring their definitions, interrelation, and philosophical implications.

Pramātā refers to the subject or the knower. It represents the conscious entity engaged in the act of knowing. In the context of perception and cognition, pramātā is the individual's subjective consciousness—the essence of awareness that enables perception, understanding, and knowledge acquisition. It is the active agent that experiences, analyses, and interprets the world of objects.

The *Cidgaganacandrika* offers an intriguing exposition on the levels of pramātā in the Krama tradition of Kashmiri Shaivism. These seven perceivers represent different states of consciousness and guide the journey back to the source of manifestation. The *sakala pramātā* operates in the realm of objectivity, experiencing the world with all three malas (impurities)—*āṇava, māyīya,* and *kārmaṇa.* These malas are obstacles that obscure true self-realisation: āṇava mala signifies the ego and sense of separation, māyīya mala embodies illusion and ignorance, and kārmaṇa mala relates to the binding effects of past actions and karma. Together, they hinder the perception of divine unity and perpetuate cycles of rebirth and suffering.

It further elaborates that the *pralayākala pramātā* exists in a state of void or unawareness, with only āṇava and māyīya malas active. The *vijñānakala pramātā*, achieved through yoga, possesses active awareness but remains unstable. The *mantra pramātā* attains a state of self-awareness, yet instability persists. The *mantreśvara pramātā* reaches a more stable unity with the universe, while the *mantramaheśvara pramātā* experiences a solid, unchanging realisation of oneness with the universe. The final state, *śiva pramātā*, is where both the state and observer merge into the pure consciousness of Śiva, embodying complete unity and transcendence.

Prameya refers to the object of knowledge or the known. It encompasses the external world, phenomena, and concepts that are the focus of the subject's awareness and understanding. Prameya includes everything that can be perceived, conceptualised, or comprehended by the pramātā, from tangible objects to abstract ideas, encompassing both physical entities and metaphysical truths.

The relationship between pramātā and prameya is intricate and interdependent. Pramātā relies on prameya for knowledge acquisition, as objects serve as stimuli for perception, cognition, and understanding. Conversely, prameya gains significance and relevance through the subject's awareness and interpretation. The subject and object coexist, each influencing and shaping the other in the process of acquiring and comprehending knowledge.

These concepts also carry significant ethical implications. Recognising the interdependence of subject and object can foster empathy, compassion, and a sense of interconnectedness. Understanding the dynamic relationship between the self and the world encourages responsible and harmonious engagement with others and the environment. Moreover, the concepts of pramātā and prameya are fundamental in epistemology, the study of knowledge and its acquisition. Understanding their relationship clarifies the nature of perception, cognition, and the validity of knowledge. Exploring how the subject interacts with the object through various cognitive processes enriches our understanding of the foundations of knowledge and truth.

Subject and Object in Indian Philosophy: Vedanta, Buddhism, and Nyaya Perspectives

In this section, we will investigate the concepts of pramātā and prameya within the frameworks of Vedanta, Buddhism, and Nyaya, revealing how each tradition navigates the complexities of consciousness, reality, and epistemology.

1. In Vedanta, pramātā is identified as the individual self or consciousness (atman), while prameya represents the phenomenal world or the universe (*jagat*). The goal of Vedanta is to realise the non-dualistic identity of pramātā (atman) with the ultimate reality (brahman). This realisation transcends the subject-object duality and leads to an experience of the underlying unity of existence. This is qualified with other modes of the connection between atman and brahman, from dualism in *Dvaita Vedanta* to qualified non-dualism in *Vishishtadvaita Vedanta*.

2. In Buddhism, pramātā and prameya are explored through the concepts of dependent origination and emptiness. Buddhist philosophy posits that both subject and object lack inherent, independent existence. It emphasises the interdependent nature of reality, questioning the separateness of subject and object and highlighting the impermanence and emptiness of phenomena.

3. Nyaya philosophy examines pramātā and prameya from an epistemological perspective. It categorises the means of knowledge (*pramāṇa*) through which the subject acquires valid cognition of the object. The

Nyaya tradition identifies various pramāṇas, such as perception, inference, comparison, testimony, and presumption, as methods to gain valid knowledge about the object.

In conclusion, the concepts of pramātā and prameya hold significant philosophical importance in understanding perception, cognition, and reality. Exploring their interrelation provides insights into the complex dynamics between the knower and the known. By transcending the limitations of subject-object duality, we gain a profound realisation of interconnectedness, unity, and the ultimate nature of reality.

Union of Anuttara, Prakasha, and Vimarsha: Exploring Trika Shaivism's Ultimate Goal

Central to Trika Shaivism are the concepts of anuttara, prakasha, and vimarsha. Let's explore each of these concepts to gain a clearer understanding of their significance and how they are pivotal for understanding spiritual awakening and self-realisation within this tradition.

1. Anuttara represents the supreme reality, the highest state of being in Trika Shaivism. It transcends all dualities, concepts, and limitations, signifying the boundless and formless essence of existence. It underlies and permeates all levels of manifestation, serving as the source and ground of everything.
2. Prakasha refers to the unchanging, ever-present light of consciousness that illuminates all experience and

reveals the diverse objects and phenomena within our field of experience. It signifies the inherent radiance of consciousness that enables perception, cognition, and understanding.
3. Vimarsha denotes the reflective self-awareness that emerges from the interaction between the subject and object. It is the faculty of consciousness that recognises and discerns the various objects and experiences in the realm of perception. It allows the subject to reflect on its own nature and experience, leading to self-realisation and understanding of its essential unity with the supreme reality.

According to Trika Shaivism, during the process of creation's expansion, the eternal anuttara, or Shiva, manifests as the primordial sound 'A.' This sound represents the foundational life force of all letter-energies (*Sakala-kala-jaala-jivana*). During this expansion, Shiva takes the form of 'ha', symbolising Shakti, the divine energy responsible for expansion. The form of 'ha' signifies the *kundalini-shakti*, the power driving the expansion. As this process continues, Shiva manifests as a dot, representing the objective phenomena (*nara Rupena*) and embodying the full scope of Shakti's manifestation, alongside Bhairava. This dot signifies the entirety of creation and its dynamic nature.

This expansion process is represented by 'Aham' or 'I', indicating the emergence of the self. Conversely, the process of contraction or return is represented by 'Maha', symbolising the withdrawal or return to the source. This concept, known as the great secret (*etad guhyam mahaguhyam*), is the essence of the universe's emergence and is also referenced in the Mahabharata (6.40.75).

व्यासप्रसादाच्छ्रुतवानेतद्गुह्यमहं परम्।
योगं योगेश्वरात्कृष्णात्साक्षात्कथयतः स्वयम्।।

As mentioned before, vimarsha plays a crucial role in self-recognition and self-realisation. As the reflective consciousness, it enables individuals to turn their awareness inward and discern their true nature. Through practices such as self-inquiry, meditation, and contemplation, practitioners can realise their fundamental identity with the supreme reality (anuttara), transcending individual limitations.

Trika Shaivism emphasises the path of recognition (pratyabhijñā) for awakening to this supreme reality. This path involves recognising the divine nature within oneself and all of existence. By embracing inherent divinity in every object and experience, individuals can transcend the illusion of separateness and realise the interconnectedness and unity of all that exists. In aphorism 8 of *Svātantryasūtravṛtti*, it is stated that the power of the Self manifests both as self-awareness (I-consciousness) and as the universe, which is an expansion of that self-awareness.

सा शक्तिरहंविमर्शभावेन तत्स्फारात्मकविश्वरूपेण वा तिष्ठति

In Trika Shaivism, various tantric practices and rituals are employed to experience anuttara (the supreme reality), prakasha (the light of consciousness), and vimarsha (reflective consciousness). These practices often utilise external objects, symbols, and rituals to invoke the divine presence and deepen one's connection with the supreme reality. Through these rituals, individuals can purify their perception, awaken the light of consciousness, and cultivate the reflective awareness necessary for self-realisation.

In the *Īśvarapratyabhijñākārikā* (2.2.4), it is explained that being firmly established in one's own Self, with a clear and distinct perception, involves experiencing unity through the singular perceiver. This state reveals an interconnected vision where the forms and sequences of things are perceived as part of a unified whole.

स्वात्मनिष्ठा विविक्ताभा भावा एकप्रमातरि।
अन्योन्यान्वयरूपैक्ययुजः सम्बन्धधीपदम्॥

This concept is further elaborated in *Īśvarapratyabhijñākārikā* (4.17), which highlights that although the fundamental reality, revered and worshipped by many, exists in its ultimate form, it remains undistinguished from the ordinary world. The true grandeur of this fundamental reality is not outwardly displayed but is realised through direct, personal experience.

तैस्तैरप्वुपयाचितैरुपनतस्तन्व्याः स्थितोऽप्यन्तिके कान्तो लोकसमान
एवमपरिज्ञातो न रन्तुं यथा।
लोकस्यैष तथानवेक्षितगुणः स्वात्मापि विश्वेश्वरो नैवालं निजवैभवाय
तदियं तत्प्रत्यभिज्ञोदिता॥

In summary, the ultimate goal in Trika Shaivism is the union of anuttara, prakasha, and vimarsha. This union signifies the realisation of the essential unity of all aspects of existence, recognising that the subject, object, and supreme reality are inherently interconnected and inseparable. Achieving this union leads to the direct experience of the divine within oneself and the realisation of one's essential nature as the supreme reality.

Chapter 5
Echoes of the Empyrean in Pratyabhijñā

*I*llusions abound, veiling our nature, our reality, and our truth. In the profound tradition of Trika Shaivism, the practice of recognition of our true self, known as pratyabhijñā, holds a central place. In this chapter, we will explore the practice of recognition in Trika Shaivism, delving into its significance, methods, and implications for spiritual awakening and transformation.

Pratyabhijñā is a spiritual path that invites individuals to awaken to the divine essence that permeates all of existence. This practice, rooted in self-recognition and self-realisation, allows practitioners to transcend limited perception and embrace their inherent divinity. Pratyabhijñā is derived from the Sanskrit term meaning 'recognition' or 'revelation'. It refers to the process of recognising and realising the divine essence within oneself and all of creation. It involves awakening to the fundamental truth that the individual consciousness (pramātā) is none other than the supreme consciousness (anuttara). It is a journey of self-discovery, where the practitioner unveils their true nature as the divine.

Self-recognition in Trika Shaivism is not a mere intellectual understanding but a direct experiential realisation. It involves recognising one's own divine essence as an inseparable part of the supreme reality. Self-recognition is the key to realising the unity of subject and object, transcending limited identity, and merging with the ultimate reality. Trika Shaivism acknowledges that individuals often perceive a sense of separation from the divine due to limited perception and conditioning. This illusory sense of separation leads to suffering and a fragmented understanding of reality. The practice of recognition seeks to dissolve this illusion and unveil the inherent unity that exists between the individual and the divine.

The practice of recognition begins with cultivating awareness, as is beautifully illustrated in the following *shloka* from the Srimad Bhagavad Gita (13.24):

य एवं वेत्ति पुरुषं प्रकृतिं च गुणैः सह।
सर्वथा वर्तमानोऽपि न स भूयोऽभिजायते॥

The knowledge Shree Krishna speaks of is not merely academic or theoretical; it is realised wisdom. This profound understanding is also discussed in *Śvetāśvataropaniṣad* (1.8):

संयुक्तमेतत् क्षरमक्षरं च व्यक्ताव्यक्तं भरते विश्वमीशः।
अनीशश्चात्मा बध्यते भोक्तृ-भावाज् ज्ञात्वा देवं मुच्यते सर्वपाशैः॥

Ignorance of the true nature of reality binds the soul, while true knowledge and wisdom free it from the fetters of maya. Through meditation, contemplation, and mindfulness practices, individuals learn to observe their thoughts, emotions, and sensory experiences. This heightened awareness attunes practitioners to the subtle realms of consciousness and the underlying unity beyond apparent duality. Trika Shaivism acknowledges the existence of veils that obscure the recognition of the divine, including ignorance, conditioning, and the illusion of separateness. The practice of recognition involves systematically removing these veils through self-inquiry, introspection, and self-reflection. By unravelling layers of conditioning and understanding the limitations of the egoic mind, individuals can pierce these veils and realise their true divine nature. Additionally, in Trika Shaivism, grace (*anugraha*) plays a crucial role in the practice of recognition. Grace is seen as a divine blessing that supports and guides practitioners along their spiritual path. It is through grace that individuals gain glimpses of their true divine nature and receive the inspiration and guidance necessary for self-realisation.

The practice of recognition in Trika Shaivism is not confined to specific techniques or rituals but embraces a holistic approach to spiritual development. It integrates practices such as mantra recitation, meditation, self-inquiry, devotional rituals, and the study of sacred texts. These practices synergistically deepen awareness, purify the mind, and facilitate direct realisation of the divine essence. Moreover, Trika Shaivism emphasises that the practice of recognition leads to profound transformation and liberation, echoing the teachings found in the *Isha Upanishad*.

यस्मिन्तसर्वाणि भूतान्यात्मैवाभूद्विजानतः।
तत्र को मोहः कः शोकं एकत्वमनुपश्यंतः॥

The verse above illustrates how a knower who discovers unity sees all beings as his/her very Self, leading to the cessation of delusion and sorrow. As individuals awaken to their true divine nature, their perception of reality shifts profoundly, i.e., the limitations of egoic identity diminish and a sense of expanded awareness and interconnectedness emerges. This transformation liberates from suffering, allows the realisation of inherent divinity, and aligns with the supreme reality. *Kaṭhopaniṣad* (1.2.18) echoes a similar sentiment, affirming that the wise atman is neither born nor does it die, nor does it originate from anywhere or become anything. It is unborn, eternal, ancient, and indestructible, transcending physical death even as the body perishes.

न जायते म्रियते वा विपश्चिन्नायं कुतश्चिन्न बभूव कश्चित्।
अजो नित्यः शाश्वतोऽयं पुराणो न हन्यते हन्यमाने शरीरे॥

Barriers to Pratyabhijñā

Pratyabhijñā, the practice of recognition, is a profound spiritual path in Trika Shaivism that invites individuals to awaken to their true divine nature. While this practice offers transformative potential, it comes with challenges. Understanding and addressing these hurdles can help individuals navigate their spiritual journey with clarity, perseverance, and success.

Ignorance and limited perception present significant obstacles on the path of pratyabhijñā. Many may be unaware of their true divine nature and caught in the illusion of separateness. Overcoming this hurdle involves self-inquiry, studying sacred texts, and seeking guidance from experienced teachers who can dispel ignorance and deepen your understanding of reality. Additionally, the ego, with its attachments and identifications, poses another major challenge to pratyabhijñā. It reinforces the illusion of individuality and fosters attachment to personal desires and identities. Overcoming this hurdle involves cultivating self-awareness, practising selflessness and detachment, and recognising the impermanent nature of the egoic self. Practices like meditation and self-inquiry can aid in unravelling egoic identification and opening up to the recognition of the divine essence.

Deeply ingrained conditioning and belief systems can also hinder progress on the path of pratyabhijñā. These patterns often shape one's worldview and limit the capacity for expanded awareness. Overcoming this obstacle requires introspection, questioning long-held beliefs, and cultivating an open mind. Challenging conditioning patterns and embracing new perspectives can pave the way for expanded awareness and the recognition of the divine. Resistance to change and fear of the unknown can also impede progress on the path of pratyabhijñā. The mind often clings to familiarity and resists stepping into unfamiliar territory. Overcoming this hurdle involves cultivating courage and embracing uncertainty. It requires a willingness to step outside comfort zones, explore new possibilities, and trust in the transformative power of the divine. Practising surrender and developing faith can help individuals navigate the fear and resistance that arise.

Therefore, consistency and discipline are essential for deepening the practice of pratyabhijñā. Without them, individuals may struggle to establish a regular spiritual practice. Overcoming this hurdle involves cultivating self-discipline, creating a structured routine, and setting clear intentions. Seeking support from a community or a spiritual mentor can also provide accountability and encouragement to maintain consistency. Furthermore, distractions and mind wandering can disrupt concentration and focus in the practice of pratyabhijñā. The mind often engages in thoughts, desires, and external stimuli. Overcoming this hurdle requires cultivating mindfulness and concentration, bringing the mind back to the present moment. Regular meditation practice, breath awareness, and techniques like mantra repetition can help develop focused attention and minimise distractions.

Attachment to specific experiences and expectations can hinder the practice as well. Individuals may become fixated on achieving particular spiritual states or experiences, leading to disappointment and frustration when expectations are not met. Overcoming this hurdle involves cultivating non-attachment and surrendering the need for specific outcomes. Embracing the present moment, being open to whatever arises, and cultivating gratitude for each experience can alleviate attachment and allow for a deeper recognition of the divine. Moreover, seeking validation and external approval can also be significant obstacles on the path of pratyabhijñā. The need for validation from others can distract individuals from their inner journey and lead to self-doubt. Overcoming this hurdle requires cultivating self-acceptance, self-compassion, and self-validation. This is echoed in Shloka 1.6 of the *Kauṣītaki Upaniṣad,* which speaks of all existence being the Self. Understanding that the recognition of the divine is an inward journey that transcends external validation allows individuals to stay focused on their spiritual path.

ऋतुरस्यार्तवोऽस्याकाशाद्योनेः सम्भूतो भार्यायै रेतः
संवत्सरस्य तेजोभूतस्य भूतस्यात्मभूतस्य त्वमात्मासि
यस्त्वमसि सोहमस्मीति तमाह कोऽहमस्मीति सत्यमिति ब्रूयात्किं
तद्यत्सत्यमिति यदन्यद्देवेभ्यश्च प्राणेभ्यश्च तत्सदथ
यद्देवाच्च प्राणाश्च तद्यं तदेतया वाचाभिव्याहियते

सत्यमित्येतावदिदं सर्वमिदं सर्वमसीत्येवैनं तदाह
तदेतच्छ्लोकेनाप्युक्तम् ॥

Overall, by addressing ignorance, transcending egoic identification, challenging conditioning, embracing change, cultivating discipline, developing concentration, and letting go of attachment and the need for external validation, individuals can navigate these challenges and make progress on the path of awakening.

How Does Pratyabhijñā Help Overcome the Subject-Object Binary?

The practice of pratyabhijñā in Trika Shaivism offers a profound path for transcending the subject-object binary, which is deeply ingrained in our perception and understanding of reality. Let's look at how the practice of pratyabhijñā seeks to overcome the subject-object binary, unveiling the deeper truth of unity and interconnectedness in Trika Shaivism.

The subject-object binary is the conventional framework through which we perceive and interpret the world. It creates a division between the observer (subject) and the observed (object), establishing a sense of separation and duality. This binary categorises our experiences, shaping our understanding of reality as fragmented and divided. Pratyabhijñā challenges the subject-object binary by emphasising non-duality. It invites individuals to recognise that the apparent division between subject and object is illusory and realise the underlying unity that transcends this binary, revealing that the subject and object are inseparably connected and interdependent. This sentiment has been succinctly and metaphorically mentioned in the *Taittirīya Upanishad* 3.9.1:

अन्नं बहु कुर्वीत। तद्व्रतम्। पृथिवी वा अन्नम्।
आकाशोऽन्नादः। पृथिव्यामाकाशः प्रतिष्ठितः।
आकाशे पृथिवी प्रतिष्ठिता।
तदेतदन्नमन्ने प्रतिष्ठितम्।
स य एतदन्नमन्ने प्रतिष्ठितं वेद प्रतितिष्ठति।
अन्नवानन्नादो भवति। महान्भवति प्रजया
पशुभिर्ब्रह्मवर्चसेन। महान् कीर्त्या॥
इति नवमोऽनुवाकः॥

Sri Adi Shankaracharya comments on this shloka, stating that in the universe, everything is interdependent; nothing is truly independent. He goes on to highlight that if anything has a dependent existence, it is an instantiation of falsity (*mithya*). This implies that creation has a borrowed existence because something else is the *adhiṣṭhānam*, the underlying reality that lends existence and substantiation to creation. And in Sri Adi Shankaracharya's view, this adhiṣṭhānam is brahman. There is also an interesting meditation on the idea that a subject cannot enjoy its subject-ness without an object, just as an object cannot enjoy its object-ness without a subject. This leads us to the conclusion that they are interdependent categories.

Another important point to note here is that Pratyabhijñā embraces paradox, recognising that reality encompasses both unity and diversity, transcendence and immanence. This practice acknowledges that while the subject-object binary serves a functional purpose in everyday perception, it does not reflect the ultimate truth of non-duality. Embracing paradox allows individuals to navigate the practical aspects of life while staying rooted in the awareness of underlying unity. This recognition deepens empathy, compassion, and interconnectedness, fostering a sense of unity and harmony with all of creation. The primacy of paradox is exemplified in

Shiva Nataraja, who embodies creation with the drum in one hand and destruction with fire in the other. His unconcerned smile contrasts with the meaningful gestures of his dance, highlighting the intricate balance of opposing forces. As physicist Brian Greene has stated:

> 'In the dance of Nataraja, we find the beauty of mathematical precision and the mystery of artistic expression converging. It is a celebration of both the rational and the intuitive aspects of human consciousness.'

Pratyabhijñā seeks to facilitate a direct experiential realisation of the unity beyond the subject-object binary. Through deep contemplation, meditation, and self-inquiry, individuals can access a state of consciousness where the boundaries between subject and object dissolve. This direct experience of unity allows for a profound understanding of the interconnectedness and unity that lie beyond the subject-object binary. By transcending the subject-object binary, individuals can perceive the divine essence in everything they encounter, as highlighted by the renowned seer, Abhinavgupta:

> 'Nothing perceived is independent of perception and perception differs not from the perceiver, therefore the universe is nothing but the perceiver.'

Abhinavagupta also proposed an intriguing method to meditate and attain non-dual awareness. This approach instructs the practitioner not to abandon anything, not to accept anything, but to pause and abide in one's Self.

मा किञ्चित्त्याज मा गृहाण।
विरम स्वस्थो यथावस्थितः॥

Here, Abhinavagupta delineates the core of *Śāmbhavopāya* as the means or method of attaining non-dual awareness centred on Śambhu or Shiva. This approach involves being present to the interplay between subject and object without becoming attached to the perception of duality. In this *upāya,* practitioners neither abandon nor accept anything. Meaning, that mind continues to engage in various thoughts, yet it does not try to control them, nor does it accept any particular thoughts; it simply remains as a witness to all experiences.

It is also crucial to realise that Pratyabhijñā extends beyond formal meditation or contemplation sessions; it permeates all aspects of life. Practitioners seek to integrate the awareness of non-duality into their everyday activities, relationships, and interactions. The *Mālinīvijayatantra,* a sacred scripture of Trika Shaivism, describes *āṇavopāya* as the method pertaining to the limited being (aṇu) in the following manner:

उच्चारकरणध्यानवर्णस्थानप्रकल्पनैः ।
यो भवेत्स समावेशः सम्यगाणव उच्यते ॥

Here, *uccāra* involves focusing attention on various aspects of vital energy (prāṇa). *Karaṇa* harnesses the physical body and senses, while *dhyāna* denotes contemplation. *Varna* pertains to listening to the subtle, unstruck sound known as *anāhata,* and *sthānakalpanā* involves fixing the mind on specific places. This integrated approach facilitates a transformative experience of unity amid diverse experiences and circumstances.

In Vijñānabhairava 52, a profound technique known as *Kālāgni* or 'the fire of time,' is described. This technique metaphorically describes the transformative process where the bodily manifestations of the individual are metaphorically burned away, leading to a profound experience of inner peace.

कालाग्निना कालपदादुत्थितेन स्वकं पुरम्।
प्लुष्टं विचिन्तयेदन्ते शान्ताभासस्तदा भवेत्॥

When discussing pratyabhijñā, it is also essential to consider the role of negation or contradiction. As highlighted in *Īśvarapratyabhijñākārikā* (2.3.15), the verse illustrates how, within the vast universe—a canvas of diverse phenomena—the essence of Shiva is realised through the interplay of opposites.

विश्ववैचित्र्यचित्रस्य समभित्तितलोपमे।
विरुद्धाभावसंस्पर्शे परमार्थसतीश्वरे॥

This perspective is crucial for understanding Shiva, who transcends all binaries. In Sanatana Dharma, the via-negativa approach is used to recognise the fundamental reality of the universe by identifying what it is not. This approach is encapsulated in the famous aphorism 'नेति नेति' (neti neti), meaning 'not this, not that'. This method of characterisation has influenced practices such as *nindāstuti* (devotion by censure) in Bharat. Since the fundamental reality of Shiva is neither a specific proposition nor its negation but exists beyond both, examining opposites and contradictions helps stretch the logical bounds and move toward recognising the transcendental truth of Shiva.

In conclusion, the practice of pratyabhijñā in Trika Shaivism provides a powerful framework for transcending the contradicting dualities of life. Through the cultivation of non-dual awareness, self-recognition as divine consciousness, and the integration of Shiva and Shakti, practitioners can overcome the limitations of the subject-object binary and experience the profound truth of the unity of all beings and phenomena

Section II

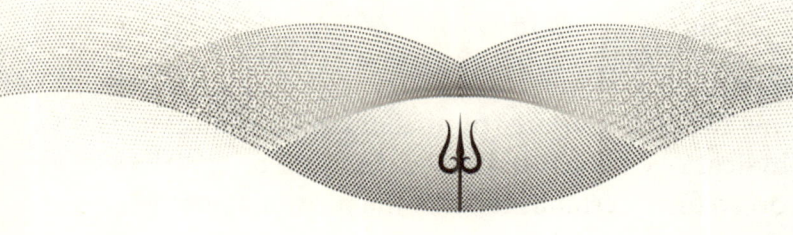

Chapter 6
The Quantum Leap Forward

The revolutionary advent of quantum mechanics in the early 20th century marked a paradigm shift in our understanding of the fundamental nature of reality. Quantum mechanics transformed our perception of the microscopic world, challenging the foundations of classical physics and introducing a new framework to describe the behaviour of particles and waves. In this chapter, we will explore the revolutionary advent of quantum mechanics, its key principles, groundbreaking discoveries, and its profound impact on science and technology and on our understanding of the universe.

A New Paradigm

At the turn of the 20th century, the scientific community found itself confronted with a crisis. The prevailing theories of that time, built upon the foundation of classical mechanics and electromagnetism, faced significant challenges and inconsistencies. This was further worsened by the arrogance of certain schools of thought regarding the comprehensiveness of the scientific constructs of the age, as seen in Lord Rayleigh's words:

'The more important fundamental laws and facts of physical science have all been discovered, and these are now so firmly established that the possibility of their ever being supplanted in consequence of new discoveries is exceedingly remote.'

Before the advent of quantum physics, classical physics, which was developed over several centuries, provided a robust understanding of the physical world. Newton's laws of motion and the mathematical elegance of classical mechanics revolutionised our perception of motion, while Maxwell's equations beautifully described electromagnetism, whose crux lies in his own words:

'We can scarcely avoid the inference that light consists in the transverse undulations of the same medium which is the cause of electric and magnetic phenomena.'

Classical physics offered predictive power and explained phenomena on a macroscopic scale. However, as scientific knowledge advanced and precision measurements became possible, certain phenomena emerged that defied classical explanations. These anomalies challenged the assumptions and predictions of classical physics. Examples include the ultraviolet catastrophe in blackbody radiation and the inability to explain the stability of atoms. The crisis in classical physics prompted scientists to seek a new framework capable of resolving the inconsistencies and explaining these novel phenomena. This marked a pivotal moment in scientific history and profoundly altered our understanding of life. Pioneers such as Max Planck, Albert Einstein, and Niels Bohr proposed groundbreaking ideas that ultimately led to the birth of quantum theory. Max Planck's work on blackbody radiation introduced the concept of energy quantisation. Planck's hypothesis that energy exists in discrete packets, or 'quanta', challenged the continuous nature of classical physics. This was the first step towards a transformative shift in our perception of the microscopic world. Einstein's explanation of the photoelectric effect in 1905 further highlighted the limitations of classical physics. He proposed that light behaves as discrete particles, later known as photons, rather than continuous waves. This wave-particle duality shattered traditional notions of light and laid the foundation for quantum mechanics. In his words, he put it this way:

'It seems as though we must use sometimes the one theory and sometimes the other, while at times we may use either. We are faced with a new kind of difficulty. We have two contradictory pictures of reality; separately neither of them fully explains the phenomena of light, but together they do.'

Niels Bohr's Copenhagen interpretation, developed in the 1920s, offered a framework for understanding quantum phenomena. It emphasised the fundamental role of observation and introduced the concept of uncertainty, stating that the precise determination of certain properties of particles is inherently limited.

All in all, the far-reaching implications of such developments cannot be overstated. It has revolutionised our understanding of fundamental particles, atomic structure, and the behaviour of matter and energy on the microscopic scale. This, in turn, has paved the way for advancements in various fields, including quantum computing, quantum cryptography, and quantum simulations. The crisis in classical physics was ultimately resolved through the acceptance and development of quantum mechanics. This new framework provided a more accurate description of physical phenomena and reconciled the inconsistencies encountered in classical theories. In Feynman's words:

> 'I am going to tell you what nature behaves like. If you will simply admit that maybe she does behave like this, you will find her a delightful, entrancing thing. Do not keep saying to yourself, if you can possibly avoid it, "But how can it be like that?" . . . Nobody knows how it can be like that.'

It's also important to note that the crisis in classical physics at the turn of the twentieth century served as a catalyst for scientific progress. The limitations and anomalies observed in classical physics not only challenged the prevailing beliefs and theories but also prompted scientists to seek a new understanding of the microscopic world, ultimately leading to the revolutionary advent of quantum mechanics. Quantum mechanics emerged as a powerful framework, providing fresh insights into the fundamental nature of reality and transforming our understanding of physics and beyond.

Ontological Departure from Classical Physics with Planck and Einstein

As discussed, the onset of quantum physics in the early 20th century, driven by the groundbreaking contributions of Max Planck and Albert Einstein, signified a profound ontological departure from classical physics. Quantum physics challenged the deterministic and reductionist worldview of classical physics. Max Planck's work on blackbody radiation in 1900 played a pivotal role in initiating the quantum revolution. Planck introduced the idea of energy quantisation, challenging the classical understanding of energy as a continuous entity. Einstein's contributions to quantum physics further propelled the ontological shift from classical physics. His explanation of the photoelectric effect, which suggested that light behaves as discrete particles, challenged the wave-particle duality of classical physics and underscored the necessity for a new framework to reconcile quantum and classical physics—two fundamentally different ways of understanding the universe. However, these discoveries, which highlighted the probabilistic nature of quantum theory, bothered Einstein and led to important debates with other physicists like Niels Bohr. On the *Bohr-Einstein debates*, Leonard Susskind once remarked:

'Albert Einstein, who was in many ways the father of quantum mechanics, had a notorious love-hate relation with the subject. His debates with Niels Bohr—Bohr completely accepting of quantum mechanics and Einstein deeply sceptical—are famous in the history of science. It was generally accepted by most physicists that Bohr won and Einstein lost. My own feeling, I think shared by a growing number of physicists, is that this attitude does not do justice to Einstein's views. Both Bohr and Einstein were subtle men. Einstein tried very hard to show that quantum mechanics was inconsistent; Bohr, however, was always able to counter his arguments. But in his final attack, Einstein pointed to something so deep, so counterintuitive, so troubling, and yet so exciting, that at the beginning of the twenty-first century, it has returned to fascinate theoretical physicists. Bohr's only answer to Einstein's last great discovery—the discovery of entanglement—was to ignore it.'

This was followed by the revolutionary work of Werner Heisenberg, Erwin Schrödinger, and others which formalised quantum mechanics. They embraced the uncertainty principle which states that there are inherent limits to simultaneously measuring certain properties of particles and the wave-particle duality which recognises that particles can exhibit both wave-like and particle-like behaviours, depending on the context of observation. These concepts challenged notions of causality and determinism, introduced probabilities, and acknowledged the interconnectedness of particles. Werner Heisenberg was among those who deliberated on the point of causality, saying:

> 'It is true that in quantum theory we cannot rely on strict causality. But by repeating the experiments many times, we can finally derive from the observations statistical distributions, and by repeating such series of experiments, we can arrive at objective statements concerning these distributions.'

Quantum mechanics continues to inspire scientists and drive ongoing research. From quantum entanglement to quantum field theory, physicists strive to deepen our understanding of the quantum world and expand the frontiers of knowledge. The quest for a unified theory of quantum gravity remains a grand challenge, with potential breakthroughs awaiting discovery. Along the way, intriguing questions have emerged, such as those posed by Stephen Hawking.:

'Even if there is only one possible unified theory, it is just a set of rules and equations. What is it that breathes fire into the equations and makes a universe for them to describe?'

As research in quantum physics continues, we remain at the threshold of further insights and discoveries, expanding our understanding of the universe and our place within it.

The Quantum Leap: Transforming Science, Technology, and Philosophy

The new paradigm brought about by quantum physics has had profound implications. It revolutionised our understanding of energy, matter, and the fundamental nature of reality and paved the way for numerous technological advancements and practical applications. Quantum mechanics underlies technologies such as transistors, lasers, and quantum computers. These innovations have transformed various fields, including communication, computing, and precision measurement.

The quantum revolution has also sparked philosophical debates. The uncertainty principle and wave-particle duality raise profound questions about the nature of reality, the role of observation, and the limits of human knowledge. It has also ignited numerous discussions and debates about the fundamental nature of reality. There are various points of interest that prompt us to reconsider our general notions of reality. These include Abner Shimony's five points implicitly supplied by quantum theory, Niels Bohr's perspective on interconnected information, the benefits of the Jamesian ontological accommodation from the Copenhagen Interpretation, and the concept of a 'closed system.' Let's examine these ideas in order to gain insights into the complexities and implications of quantum theory.

Abner Shimony, a prominent physicist and philosopher, identified five fundamental points that quantum theory implicitly supplies. These points include non-locality, holistic interconnectedness, contextuality, the indeterminacy of measurement outcomes, and the participatory role of the observer. These aspects reveal the unique characteristics of quantum phenomena, inviting us to reconsider our understanding of the fabric of the universe. He highlights some of these aspects in his 1988 writing for *Scientific American*:

> 'First, two entities separated by many meters and possessing no mechanism for communicating with each other nonetheless can be "entangled": they can exhibit striking correlations in their behaviour, so that a measurement done on one of the entities seems instantaneously to affect the result of a measurement on the other. The finding cannot be explained from a classical point of view, but it agrees completely with quantum mechanics. Second, a photon, the fundamental unit of light, can behave like either a particle or a wave, and it can exist in an ambiguous state until a measurement is made. If a particle-like property is measured, the photon behaves like a particle, and if a wavelike property is measured, the photon behaves like a wave. Whether the photon is wave- or particle-like is indefinite until the experimental arrangement is specified. Finally, the notion of indefiniteness is no longer confined to the atomic and subatomic domains. Investigators have found that a macroscopic system can under some circumstances

exist in a state in which a macroscopic observable has an indefinite value. Each of these findings alters drastically the way we perceive the world.'

Niels Bohr, one of the founding figures of quantum theory, emphasised the interconnected nature of quantum systems. According to Bohr, nature's most basic element is not particles or waves but interconnected information. He argued that the properties of quantum entities are defined through their interactions and relationships with other entities, highlighting the relational nature of quantum reality. An insightful observation by Bohr, which goes beyond the empirical into the metaphysical, is the futility of division not just of entities but also the very metaphysical framework—be it of objectivity or subjectivity.

'I myself find the division of the world into an objective and a subjective side much too arbitrary. The fact that religions through the ages have spoken in images, parables, and paradoxes means simply that there are no other ways of grasping the reality to which they refer. But that does not mean that it is not a genuine reality. And splitting this reality into an objective and a. subjective side won't get us very far.'

The Copenhagen Interpretation, spearheaded by Bohr, introduced what can be regarded as the concept of 'ontological accommodation.' William James's philosophical ideas influenced this perspective, emphasising the pragmatic accommodation of different viewpoints, which is dramatically expressed in his words:

'Whenever two people meet, there are really six people present. There is each man as he sees himself, each man as the other person sees him, and each man as he really is.'

Niels Bohr introduced a similar idea, where the observer 'creates' a contextualised and relational reality through the act of observation. He explained:

'When we measure something, we are forcing an undetermined, undefined world to assume an experimental value. We are not measuring the world, we are creating it.'

In the context of quantum theory, ontological accommodation allows for the coexistence of various interpretations, recognising their complementary and context-dependent nature. In the context of quantum theory, the concept of a 'closed system' refers to a system that is isolated from external influences. It is a fundamental assumption in quantum mechanics, allowing for the application of mathematical formalisms and the prediction of measurement outcomes. The idea of a closed system enables the study and understanding of quantum phenomena within controlled experimental conditions, albeit with caveats. In his paper titled *The Quantum Postulate and the Recent Development of Atomic Theory* published in a supplement to *Nature* (14 April 1928), Bohr stated:

'Schrödinger has expressed the hope that the development of the wave theory will eventually remove the irrational element expressed by the quantum postulate and open the way for a complete description of atomic phenomena along the lines of the classical theories. In support of this view, Schrödinger, in a recent paper (Ann. d. Phys., 83, p. 956; 1927), emphasises the fact that the discontinuous exchange of energy between atoms required by the quantum postulate, from the point of view of the wave theory, is replaced by a simple resonance phenomenon. In particular, the idea of individual stationary states would be an illusion and its applicability only an illustration of the resonance mentioned. It must be kept in mind, however, that just in the resonance problem mentioned we are concerned with a closed system which, according to the view presented here, is not accessible to observation.

An important point to note here is that no system that is probed by a measuring device is entirely 'closed.' We have distinct constraints and conservation laws when we consider quantum mechanics in closed systems. In a manner of speaking, quantum mechanics tries to predict the probabilities of alternative coarse-grained time histories of a closed system. More recently, the programme to apply quantum mechanics to cosmology has raised pivotal and fundamental questions such as how do we assign probabilities in a closed quantum system corresponding to a quantum gravitational spacetime?

The discussions surrounding these aspects of quantum theory play a crucial role in shaping our view of reality. They challenge classical determinism and reductionism, highlighting the contextual and participatory nature of quantum phenomena. These discussions push the boundaries of our philosophical and scientific frameworks, stimulating further inquiry into the fundamental nature of the universe. And by engaging in these discussions, we can uncover the intricate interconnectedness, contextual nature, and participatory role of observers in the quantum realm.

Superposition and Uncertainty

Two fundamental concepts that lie at the heart of quantum theory are quantum superposition and uncertainty. In this section, we will delve into the premise of these phenomena, examining their implications for our understanding of the quantum world and the nature of existence itself.

The quantum superposition principle states that a quantum system can exist in multiple states simultaneously until it is measured or observed. This implies that particles can exist in a state of indeterminacy, occupying a combination of different states rather than being restricted to a single value. In other words, it asserts that reality is probabilistic and exists as a range of potential states until observation collapses the wave function into a specific outcome. Roger Penrose encapsulated the idea as follows:

> 'We cannot say, in familiar everyday terms, what it "means" for an electron to be in a state of superposition of two places at once, with complex-number weighting factors w and z. We must, for the moment, simply accept that this is indeed the kind of description that we have to adopt for quantum-level systems. Such superpositions constitute an important part of the actual construction of our microworld, as has now been revealed to us by nature. It is just a fact that we appear to find that the quantum-level world actually behaves in this unfamiliar and mysterious way. The descriptions are perfectly clear cut, and they provide us with a micro-world that evolves according to a description that is indeed mathematically precise and, moreover, completely deterministic!'

Recently, it has also been proposed that indefinite causal structures, which emerge from quantum superpositions of different space-time geometries, can serve as a universal resource for performing any quantum operation on spatially distributed systems. This approach allows tasks that would otherwise be impossible with only local operations and classical communication (LOCC), such as perfect teleportation and local discrimination of Bell states, to be achieved. In quantum gravity, fundamental indefiniteness in the causal structure of space-time arises because quantum superpositions of various space-time geometries make the causal structure both dynamic and indefinite. This revolutionary idea suggests that the causal order between events in space-time can also become indefinite, challenging the conventional understanding of causality.

Thus, we can see that quantum superposition challenges the notion of a fixed, objective reality prior to measurement and raises intriguing questions about the nature of reality. Adding to this complexity, the uncertainty principle, formulated by Werner Heisenberg, states that certain pairs of physical properties, such as position and momentum, cannot be simultaneously known with precise accuracy. This principle reveals an inherent limit to our knowledge of quantum systems and introduces an element of uncertainty into measurements. As of today, many quantum gravity models suggest that a minimum length emerges at the Planck scale, leading to modifications of the Heisenberg uncertainty principle, resulting in what is known as the generalised uncertainty principle.

The ontological understanding of uncertainty is closely linked to the complementary nature of observables in quantum mechanics. Certain properties of particles, such as position and momentum or energy and time, are complementary and cannot be precisely measured simultaneously. The uncertainty principle recognises this inherent trade-off between the precision of measurements. On this subject, Milton Friedman once said:

> 'In both social and natural sciences, the body of positive knowledge grows by the failure of a tentative hypothesis to predict phenomena the hypothesis professes to explain; by the patching up of that hypothesis until someone suggests a new hypothesis that more elegantly or simply embodies the troublesome phenomena, and so on ad infinitum. In both, experiment is sometimes possible, sometimes not (witness meteorology). In both, no experiment is ever completely controlled, and experience often offers evidence that is the equivalent of a controlled experiment. In both, there is no way to have a self-contained closed system or to avoid interaction between the observer and the observed. The Gödel theorem in mathematics, the Heisenberg uncertainty principle in physics, the self-fulfilling or self-defeating prophecy in the social sciences all exemplify these limitations.'

To summarise, the concepts of quantum superposition and the uncertainty principle raise profound questions about the nature of existence, the role of observation, and the limits of human knowledge. These ideas challenge our classical perceptions and suggest that reality may be far more complex and interconnected than previously thought. As research progresses, we may uncover even deeper insights into the mysteries of the quantum world.

Quantum Origins: How Tiny Fluctuations Shaped the Universe

In the early moments of the universe, during the period known as cosmic inflation, a remarkable process unfolded that laid the foundation for the formation of galaxies, stars, and ultimately, life itself. At this primordial stage, quantum fluctuations played a pivotal role in shaping the structure of the cosmos. In this section, we will explore the fascinating concept of quantum fluctuations in the early universe and their profound impact on the evolution of cosmic structures.

Quantum fluctuations arise from the inherent uncertainty principle of quantum mechanics, which posits that even in a vacuum, particles and fields undergo spontaneous fluctuations. These fluctuations occur on extremely small scales and have fleeting durations. However, during cosmic inflation, the expansion of the universe was so rapid that these tiny quantum fluctuations were stretched to cosmic scales, leaving an indelible imprint on the fabric of the universe. Sten F. Odenwald elaborated on this matter as follows:

> 'Quantum fluctuations are, at their root, completely a-causal, in the sense that cause and effect and ordering of events in time is not a part of how these fluctuations work. Because of this, there seem not to be any correlations built into these kinds of fluctuations because "law" as we understand the term requires some kind of cause-and-effect structure to pre-exist. Quantum fluctuations can precede physical law, but it seems that the converse is not true. So, in the big bang, the establishment of "law" came after the event itself, but of course, even the concept of time and causality may not have been quite the same back then as they are now.'

These quantum fluctuations served as the seeds for the formation of cosmic structures. The study of the cosmic microwave background radiation, the residual radiation from the early universe, provides compelling results regarding quantum fluctuations. The observations revealed subtle temperature variations across the sky, which correspond to the density fluctuations present during the early stages of the universe. Over time, gravity acted on these variations, causing denser regions to attract more matter and grow in size, eventually leading to the formation of galaxies, clusters of galaxies, and large-scale cosmic structures. Therefore, these temperature variations offer a window into the primordial quantum fluctuations that eventually gave rise to the large-scale structures we observe today.

Understanding the role of quantum fluctuations in the early universe is not only crucial for tracing the origins of cosmic structures but also for comprehending the fundamental nature of the universe itself. It provides insights into the interplay between quantum mechanics and gravity on cosmic scales as well as exemplifies the profound interconnections between the microscopic world of quantum mechanics and the macroscopic universe we observe today. They highlight the remarkable capacity of tiny quantum fluctuations to shape the vast cosmos. Through ongoing observations, experiments, and theoretical advancements, scientists continue to unravel the mysteries of these fluctuations, deepening our understanding of the origins and evolution of our universe.

Chapter 7
Spooky Action at a Distance

Quantum entanglement is a phenomenon where particles become interconnected, such that the state of one particle instantaneously influences the state of another, regardless of the distance between them. Described by Albert Einstein as 'spooky action at a distance,' entanglement reveals a profound interconnectedness that transcends the boundaries of space and time.

The concept of entanglement was a subject of intense debate when it was first proposed in the early 20th century. Einstein, in particular, was sceptical of this phenomenon, believing it violated his principle that nothing could travel faster than the speed of light. However, contemporary physicists, like Michio Kaku, have clarified that while entanglement involves instantaneous correlations, it does not entail any faster-than-light communication, thus preserving the integrity of relativity. He explained it as follows:

> 'Quantum entanglement allows you to send information faster than light, which upset Einstein. But Einstein has the last laugh. The information you send on quantum entanglement is random, useless information.'

Einstein believed that 'hidden variables' exist in the quantum realm and that our lack of awareness of these variables makes quantum dynamics seem probabilistic. This view contrasted sharply with the Copenhagen interpretation of quantum mechanics, which accepts the inherent probabilistic nature of quantum phenomena without invoking additional variables to restore determinism.

A pivotal experiment challenging the local hidden variable theory was the Bell test. This experiment measures the properties of entangled particles, such as their spin or polarisation, in different directions. The results of these measurements have consistently violated the predictions of local hidden variable theories, thereby supporting the Copenhagen interpretation and reinforcing the reality of quantum entanglement.

Thus, we can see that the implications of entanglement are far-reaching. Firstly, it challenges the notion of local realism, which proposes that physical properties have definite values independent of measurement and that causality is localised, meaning information cannot travel faster than the speed of light. As Kaku noted earlier, we must now qualify this by acknowledging that information transfer is subject to temporal constraints. This clarification is crucial when discussing teleportation, where the remote reconstruction of a state using an entangled channel and Bell-state measurements, supplemented by unitary transformations, adheres to relativistic principles. While information transfer may occur instantaneously, the actual process, including unitary transformations, does not surpass the speed of light. Entanglement implies a non-local correlation, where the state of one particle is entangled with another, regardless of their distance apart.

Secondly, entanglement has profound implications for information transfer. The intertwined states of entangled particles mean that manipulating the state of one particle instantaneously affects the state of the others, no matter their physical separation. This phenomenon is pivotal in quantum communication, cryptography, and quantum computing, where entanglement enables secure communication and enhances computational power. Customised projective measurements of one subsystem in an entangled system often yield intriguing quantum multipartite states in the other subsystem(s).

The concept of entanglement also poses philosophical questions about the nature of reality. Entanglement challenges our classical understanding of cause and effect by demonstrating instantaneous correlations that appear to transcend the constraints of space and time. This phenomenon prompts us to question the fundamental foundations of our reality and explore the intricate interconnections that define the universe. Despite extensive experimental verification, many aspects of entanglement's mechanisms and implications remain elusive. Ongoing research aims to deepen our understanding of entangled states, investigate entanglement across larger scales and more complex systems, and uncover the mechanisms behind its apparent non-locality. Recent developments include the generation of entanglement between macroscopic elements and studies on entanglement entropy involving gravitons and photons. These advancements highlight entanglement's role in sparking investigations into potential connections with other areas of physics, such as gravity and the nature of spacetime.

The pursuit of a theory of quantum gravity aims to reconcile quantum mechanics with Einstein's theory of general relativity, and entanglement offers tantalising clues for potential connections between these two foundational theories. Entanglement stands as one of the most perplexing and captivating phenomena in the realm of quantum mechanics. Its non-local correlations challenge classical intuitions, while promising significant technological applications and deeper insights into the nature of reality. As scientists continue to unravel the mysteries of entanglement, we embark on a fascinating journey to comprehend the profound interconnectedness that underlies the fabric of the universe.

Quantum Correlations and Entanglement: Application and Implications

In the realm of quantum mechanics, the fundamental concepts of quantum correlations and entanglement play a crucial role in shaping our understanding of the microscopic world. These phenomena reveal profound interconnections and the intricate nature of quantum systems. Let's delve deeper into both of these phenomena.

Quantum correlations refer to the relationships and connections observed between quantum systems. Unlike classical correlations, which can be explained by local interactions and predetermined properties, quantum correlations exhibit unique characteristics that challenge the classical model. The idiosyncrasies of these correlations emerge due to the intrinsic probabilistic nature of quantum mechanics, where the state of one particle is not predetermined but is entangled with the state of another. Thus, entanglement is a specific kind of quantum correlation, which has a super-set, in a manner of speaking in quantum discord. Recently, there have been even further extensions of the possible correlations beyond even quantum discord.

The role of quantum correlations and entanglement is manifold. Quantum correlations challenge the classical notion of local realism, which assumes that particles have pre-existing properties that determine their behaviour. The violation of Bell's inequalities, demonstrated in experiments, showcases the non-local nature of quantum correlations, revealing the departure from classical expectations. More importantly, quantum correlations and entanglement serve as the backbone for quantum information processing. These phenomena enable secure quantum communication, quantum cryptography, and the potential for exponential computational power through quantum algorithms. Quantum computers can harness the power of entanglement to perform parallel computations and solve complex problems efficiently. Furthermore, entanglement allows for quantum teleportation, a process in which the exact state of one particle can be transmitted to another distant particle instantaneously, without physically traversing the space in between. This phenomenon has implications for secure communication and the transmission of quantum states across large distances.

Quantum correlations and entanglement have transformed precision measurement techniques by enabling greater accuracy and sensitivity that surpass classical limits. Quantum metrology leverages entanglement to achieve unprecedented levels of precision in applications such as gravitational wave detection, quantum clocks, and magnetic field sensing. For example, optical atomic clocks face fundamental constraints, including a noise floor influenced by quantum projection noise (QPN). This noise stems from the quantum properties of the atoms and the spin-statistics of individual qubits. Although top atomic clocks are approaching this fundamental limit, it can be mitigated through quantum entanglement, which enhances stability. Recently, techniques like spin squeezing have been introduced to generate quantum entanglement, leading to significant improvements in clock performance. By using multi-particle entanglement, measurement precision can exceed traditional shot-noise limits, pushing forward advancements in quantum metrology.

All in all, quantum correlations and entanglement are subjects of intense research and exploration, especially within and across different degrees of freedom. Scientists strive to deepen our understanding of these phenomena, uncovering their underlying mechanisms for effective deployment in various applications. Hyperentanglement and hybrid entanglement have also been used in quantum networks and quantum key distribution. All in all, the role of quantum correlations and entanglement is fundamental to our understanding of quantum mechanics and its applications. They challenge classical intuitions, revealing the complexities of quantum systems and connections that surpass classical boundaries. These phenomena have practical implications for quantum information processing, secure communication, metrology, and sensing. Moreover, they drive ongoing research, stimulating investigations into the fundamentals of quantum theory and its connections to other areas of physics. By unravelling the mysteries of quantum correlations and entanglement, scientists continue to deepen our understanding of the quantum realm and expand the frontiers of human knowledge.

Einstein-Podolsky-Rosen (EPR) Paradox and Bell's Theorem

In the realm of quantum mechanics, the Einstein-Podolsky-Rosen (EPR) paradox and Bell's theorem have posed profound challenges to our understanding of the nature of reality and the fundamental principles of quantum theory. These concepts, intertwined and complementary, have sparked extensive debate and experimentation, revealing the intricate and sometimes counterintuitive aspects of quantum mechanics. The EPR paradox was proposed by Albert Einstein, Boris Podolsky, and Nathan Rosen in 1935, and it questions the completeness of quantum mechanics and challenges the concept of local realism. The paradox begins with the idea that two particles, in an entangled state, can exhibit instantaneous correlations in their properties, even when separated by vast distances.

So, if nature respects causality and locality, how can particles instantaneously affect each other? Didn't Einstein assert that the speed of light is the upper limit for any physical entity? This question arises from the phenomenon of quantum entanglement, which seems to defy these principles. Einstein famously critiqued quantum entanglement as an incomplete theory, suggesting the existence of hidden variables that would adhere to local realism and provide a more complete understanding of physical reality. In 1950, physicists Chien-Shiung Wu and Irving Shaknov observed unexpected correlations in photon pairs, unknowingly providing the first evidence of quantum entanglement. Later, David Bohm recognised the significance of their discovery, which helped shift entanglement from a theoretical concept to an area of experimental research.

Moreover, Bell's theorem, formulated by physicist John Bell in 1964, established a mathematical framework to test the predictions of quantum mechanics against local realistic theories. It presented a way to experimentally verify the existence of non-local correlations and to distinguish between quantum mechanics and local hidden variable theories.

The essence of Bell's theorem lies in the analysis of correlations between measurements made on entangled particles., demonstrating that certain statistical predictions of quantum mechanics cannot be reproduced by any local realistic theory. In other words, there are limits to the degree of correlation that can be achieved in a classical world, whereas quantum mechanics allows for correlations that violate these limits. Experimental tests based on Bell's theorem have consistently confirmed the predictions of quantum mechanics and revealed violations of local realism. These experiments demonstrated that entangled particles exhibit correlations that cannot be explained by any classical mechanism, strengthening the case for the validity of quantum mechanics and challenging the notion of local realism.

Bell's theorem also introduced several intriguing loopholes that have captured the attention of physicists and philosophers. There are three prominent loopholes associated with Bell's theorem:

1. **The Detection/Efficiency Loophole**

This loophole arises due to the possibility of imperfect detection efficiency in experiments. It suggests that if a fraction of entangled particles is not detected or measured accurately, and this can introduce biases that may allow for alternative explanations consistent with local realism. To address this loophole, experiments must strive for high detection efficiencies and minimise the possibility of measurement errors.

2. The Locality Loophole

This loophole emerges from the potential influence of hidden variables that could propagate faster than the speed of light, violating the principle of locality. If information or causal influences can travel faster than the speed of light, it could compromise the non-local correlations observed in entangled particle experiments. Experiments aiming to close this loophole must ensure that measurements are spacelike separated to prevent any superluminal communication.

3. The Freedom-of-Choice Loophole

The freedom-of-choice loophole concerns the potential for hidden variables to influence the choices made by experimenters during the setup of the experiment. If these hidden variables could manipulate the choice of measurement settings, they could potentially introduce biases that mimic the predictions of quantum mechanics. To address this loophole, experiments must ensure that measurement settings are chosen randomly and independently of any hidden variables.

Closing these loopholes in experimental tests of Bell's theorem is a formidable task, requiring meticulous design and execution. Nevertheless, significant progress has been made in recent years, with experiments aiming to minimise the influence of these loopholes and strengthen the case for quantum mechanics. Achieving closure on these fronts holds profound implications for our understanding of quantum mechanics and the nature of reality. For instance, closing the communication or locality loophole would offer definitive proof of local realism's violation. This would solidify the non-local correlations observed in entangled particle experiments, challenging classical notions of causality and locality. Simultaneously, closing the detection and freedom-of-choice loopholes would affirm that quantum mechanics' statistical predictions defy any local realistic theory. This validation would underscore the probabilistic nature of quantum mechanics and affirm the existence of non-local correlations.

Overall, the closure of Bell's loopholes not only advances fundamental science but also enhances practical applications in quantum information processing, communication, and cryptography. It would bolster confidence in the security and reliability of quantum technologies that exploit entanglement, paving the way for future advancements in these fields.

These challenges posed by Bell's loopholes drive ongoing scientific inquiry. Their resolution promises to illuminate further the mysteries of quantum mechanics and deepen our understanding of the universe.

The Concept of Non-Locality and Instantaneous Correlations

Non-locality refers to the instantaneous influence or correlation between quantum systems that are separated by vast distances. In classical physics, the notion of locality suggests that physical interactions can only occur within a finite region of space and time, propagating at a limited speed, such as the speed of light. However, quantum mechanics challenges this notion by allowing for entangled particles to exhibit non-local correlations that transcend the confines of classical locality.

Instantaneous correlations arise from the entanglement of particles, where the state of one particle is inextricably linked to the state of another, regardless of their spatial separation, like for entanglement-type transitions. When two entangled particles are measured, the outcomes of their measurements are found to be correlated instantaneously, defying our classical understanding of cause and effect. This instantaneous correlation implies a deeper interconnectedness that transcends the constraints of space and time.

The metaphysical implications of quantum non-locality and instantaneous correlations are thought-provoking. Quantum non-locality raises intriguing questions about the nature of space and time. It challenges the traditional understanding of space as a container where objects and interactions are localised and suggests a more intricate and interconnected fabric of reality, prompting us to reconsider our notions of spatial and temporal boundaries. Michio Kaku beautifully encapsulated this idea as follows:

> 'There is a cosmic "entanglement" between every atom of our body and atoms that are light-years distant. Since all matter came from a single explosion, the big bang, in some sense the atoms of our body are linked with some atoms on the other side of the universe in some kind of cosmic quantum web. Entangled particles are somewhat like twins still joined by an umbilical cord (their wave function) which can be light-years across. What happens to one member automatically affects the other, and hence knowledge concerning one particle can instantly reveal knowledge about its pair. Entangled pairs act as if they were a single object, although they may be separated by a large distance.'

The metaphysical implications of non-locality intersect with discussions of mind-body interaction. Some philosophers and scientists are also exploring how non-locality may illuminate the relationship between consciousness and the physical world. The instantaneous correlations observed in quantum entanglement also raise intriguing questions about the role of consciousness, particularly if we consider perspectives beyond physicalist ideas and beyond the concept of decoherence.

Quantum non-locality challenges our ontological assumptions about reality, suggesting a universe characterised by profound interconnectedness rather than separateness. This perspective challenges traditional views of a fragmented reality, advocating for a more holistic understanding. These metaphysical implications have sparked diverse philosophical interpretations of quantum mechanics. The many-worlds interpretation posits parallel universes to explain non-local correlations, while the Transactional Interpretation suggests a timeless exchange of information across time. While these interpretations prompt deep philosophical inquiries, it is crucial to note ongoing debates and evolving perspectives in both philosophy and science. These discussions enrich our understanding of reality, offering new insights into the nature of space, time, causality, and the interconnected quantum realm.

As we delve into the metaphysical landscape of quantum mechanics, we confront profound questions that challenge classical intuitions and invite exploration into the mysteries at the core of this captivating field.

Measurement and Orchestration by Entanglement

In the realm of quantum mechanics, measurement is a fundamental process that reveals information about the properties and behaviour of quantum systems. This section explores the connection between quantum measurement, quantum correlations, and entanglement, shedding light on the complex nature of the quantum realm.

When a measurement is conducted on an entangled system, it unveils correlations between the properties of the entangled particles. This measurement is crucial for observing the non-local characteristics of entanglement and confirming the predictions of quantum mechanics. The measurement of entangled particles exhibits unique characteristics that distinguish it from classical measurements. In classical physics, measurements reveal the pre-existing properties of a system with certainty. However, in the quantum world, measurements yield probabilistic outcomes due to the fundamental uncertainty associated with quantum states. The act of measurement causes the quantum state of the system to 'collapse' into one of its possible states, randomly selecting a specific outcome.

Quantum measurement in the context of entanglement allows us to observe the non-local correlations between entangled particles, and even at times to 'activate entanglement'. For example, consider a pair of entangled photons prepared in a Bell state. When a measurement is performed on one photon, its state collapses, and the state of the other photon becomes instantaneously determined, regardless of the spatial separation between them. This measurement process demonstrates the non-local nature of entanglement and confirms the existence of quantum correlations.

Furthermore, quantum measurement is essential for quantifying and characterising quantum correlations and entanglement. Various measures, such as entanglement entropy, concurrence, or Bell inequalities, are used to quantify the degree of entanglement and the strength of quantum correlations between entangled systems. These measures rely on the outcomes of quantum measurements to assess the level of correlation and entanglement present in the system.

The act of measurement can also generate entanglement between the measured system and the measurement apparatus, which is the primary reason for decoherence in various systems. This occurs when the measurement process involves a quantum state, and the interaction between the system and the apparatus leads to an entangled state between them. In a process known as 'projective measurement', a measurement performed on a subsystem of a quantum system can generate entanglement between that subsystem and the rest of the system. The generation of entanglement through quantum measurement highlights the remarkable interplay between measurement, quantum states, and entanglement. It demonstrates how the act of observing or extracting information about a quantum system can have profound effects, transforming the state of the system and creating entangled relationships.

Quantum measurements can generate entanglement in various ways, particularly through techniques such as stroboscopic quantum non-demolition measurements in atomic ensembles. Research has demonstrated that measurements can induce phase transitions in quantum systems, resulting in different regimes. In some cases, interactions dominate, and entanglement becomes widespread, while in others, measurements can suppress entanglement. However, when a quantum system becomes entangled with its environment, it can result in a loss of quantum-ness, depleting the system's coherent quantum properties. This leads to environment-induced superselection, known as 'einselection', which involves selective information loss. Einselected pointer states remain stable and can maintain correlations with the broader universe despite environmental influences. Einselection also essentially restricts the use of most of the Hilbert space, promoting classical behaviour.

Understanding the generation of entanglement through quantum measurement is not only of fundamental significance but also has practical implications. It forms the basis for various applications in quantum communication, quantum computation, and quantum information processing. By exploiting the ability of quantum measurement to generate entanglement, scientists and engineers can develop protocols and technologies that harness the power of entanglement for secure communication, enhanced computing capabilities, and improved precision measurements.

Applications of Entanglement

Quantum entanglement has captivated scientists and researchers for decades. While entanglement challenges our classical understanding, it also holds great promise for a wide range of practical applications. In this section, we will explore some of the fascinating applications of quantum entanglement and the transformative impact it can have in various fields.

1. Secure Information Sharing

One of the most prominent applications of entanglement is in quantum communication, particularly in the field of quantum cryptography. Entangled particles can be used to establish secure communication channels that are inherently resistant to eavesdropping. Through protocols such as quantum key distribution (QKD), information can be shared between two parties with the guarantee of confidentiality, as any attempt to intercept the communication will disrupt the entanglement and be detectable.

Recently, a research team achieved continuous 24-hour quantum key distribution (QKD) over a 20-kilometer terrestrial free-space channel. They accomplished this by developing a robust 625-MHz decoy-state light source and implementing daytime noise suppression techniques that approached the Fourier transform limit. This setup resulted in an average secure key rate of approximately 495 bits per second (bps). In addition, the team integrated bidirectional laser communication into both the QKD transmitter and the ground station, allowing for real-time key distillation and reducing the processing time from days to real-time. This advancement lays a solid foundation for implementing continuous, real-time QKD with quantum satellites.

2. Improved Computational Speed

Entanglement lies at the heart of quantum computing, a revolutionary paradigm that promises to solve complex problems exponentially faster than classical computers. Quantum bits, or qubits, can be entangled to encode and process information in parallel, enabling quantum algorithms to outperform classical algorithms in specific applications. Entanglement is helpful when performing quantum operations and enabling quantum coherence, which is crucial for the computational power of quantum computers.

Google recently introduced a quantum computer that utilises quantum annealing technology, boasting a processing speed up to 100 million times faster than any classical computer in its lab. In our data-driven era, where we generate 2.5 exabytes of data daily—equivalent to the storage capacity of 5 million laptops—quantum computers promise to manage this immense volume of information more efficiently. They have the potential to process large datasets more effectively while significantly reducing power consumption by a factor of 100 to 1,000.

3. High-Precision Measurements

Entanglement has significant implications for high-precision measurements and sensing. By utilising entangled states, quantum sensors can achieve superior sensitivity and accuracy compared to their classical counterparts. This is particularly valuable in fields such as gravitational wave detection, magnetic field sensing, and atomic clocks. Entanglement-enhanced measurements enable more precise measurements, enabling advancements in fundamental research and practical applications.

Entanglement-based techniques are advancing imaging and microscopy by enhancing resolution and detail. For instance, quantum imaging schemes that use entangled photon pairs can surpass the diffraction limit, leading to sharper and more detailed images. Quantum microscopy, utilising these entangled photons, offers high-precision measurements with reduced noise, enablingthe visualisation of nanoscale structures and delicate biological systems with greater clarity.

4. Advanced Quantum Simulations

Entanglement plays a crucial role in quantum simulation, where quantum systems are used to simulate and study complex physical phenomena that are challenging to analyse with classical computers. By engineering entangled states, scientists can simulate the behaviour of quantum systems, helping to uncover new insights into materials, chemical reactions, and quantum phenomena.

Simulating quantum many-body systems is a key application for next-generation quantum processors. While analogue quantum simulation has demonstrated its potential for achieving quantum advantage, recent research has increasingly focused on digital quantum simulation. This shift is due to advancements in devices aimed at achieving general-purpose quantum computing capabilities.

5. Biological Research

The role of entanglement in biological systems is an area of active research. It is hypothesised that entanglement could play a role in quantum processes occurring in biological systems, such as photosynthesis and avian navigation. Understanding and harnessing entanglement in biological systems may open new avenues for advancements in medicine, biochemistry, and biotechnology.

Recently, it was hypothesised that the rapid and precise binding of certain molecules to short DNA sequences might involve quantum effects, specifically $\pi-\pi$ electron entanglement. Using restriction enzymes EcoRV and EcoRI as models, we propose that entanglement between π–π electrons at DNA-protein interfaces aids in nucleotide recognition. Additionally, phosphorescence from triplet states could potentially provide experimental evidence for this phenomenon.

These are just a few examples of the vast potential and applications of quantum entanglement, demonstrating its ability to catalyse revolutionary advancements across a spectrum of scientific and technological disciplines. While many practical applications are still in their early stages, ongoing research and advancements in the field of quantum entanglement continue to push the boundaries of what is possible. As scientists deepen their understanding of entanglement and develop innovative techniques for harnessing and manipulating it, we can expect even more groundbreaking applications to emerge in the future.

Chapter 8
Fleeting Quanta and Fading Waves

Many luminaries have likened understanding the quantum realm to chasing a will-o'-the-wisp. The epistemological challenges are vast, and even from an ontological perspective, grasping the nature of quantum phenomena is elusive, akin to trying to hold sand. In fact, it is far more delicate and subtle than sand, making it a profoundly complex area of exploration.

This complexity is highlighted by concepts such as decoherence and quantum dissipation, which are fundamental to understanding how the quantum world transitions to the classical realm. These phenomena shed light on the processes that lead to the loss of quantum coherence and the emergence of classical behaviour in macroscopic systems. In this section, we delve into the intricate nature of decoherence and quantum dissipation, exploring their mechanisms, implications, and the challenges they pose in the study of quantum systems.

Quantum Dissipation and Decoherence

When a quantum system interacts with its surrounding environment, it becomes entangled with the environmental degrees of freedom. These interactions cause the quantum superposition states to lose their delicate phase relationships, leading to a phenomenon known as decoherence. The primary cause of decoherence is the interaction of a quantum system with its surrounding environment, which can be composed of particles, fields, or other degrees of freedom. Entanglement between the system and the environment results in the spread of information and the destruction of delicate quantum interference patterns. As a result, the system appears to exhibit classical behaviour, with distinct, well-defined states rather than existing in a superposition of states. Writer Philip Ball eloquently phrases it as follows:

> 'Decoherence is what destroys the possibility of observing macroscopic superpositions—including Schrödinger's live/dead cat. And this has nothing to do with observation in the normal sense: we don't need a conscious mind to "look" in order to "collapse the wave function". All we need is for the environment to disperse the quantum coherence. This happens with extraordinary efficiency—it's probably the most efficient process known to science. And it is very clear why size matters here: there is simply more interaction with the environment, and therefore faster decoherence, for larger objects.'

Decoherence has significant implications for the transition from the microscopic quantum world to the macroscopic classical world. It provides an explanation for the apparent collapse of the quantum wave function into a single state when observed macroscopically and explains why macroscopic objects, such as everyday objects, appear to behave classically rather than exhibiting the bizarre and counterintuitive quantum behaviour of superposition and entanglement.

Quantum dissipation, on the other hand, focuses on the loss of energy and coherence in quantum systems due to interactions with their environments. These interactions lead to the dissipation of energy and the degradation of quantum states. Quantum dissipation is particularly relevant in the study of open quantum systems, where the system of interest interacts with an external environment that acts as a dissipative agent. In open quantum systems, the environment causes the loss of energy and the decay of quantum states through processes such as spontaneous emission, scattering, or energy exchange. The dissipation of energy to the environment leads to the emergence of classical behaviour, where the system thermalises and reaches a state of equilibrium.

Understanding and characterising the mechanisms of decoherence and quantum dissipation pose significant challenges in the study of quantum systems. These phenomena are influenced by various factors, including the strength and nature of the system-environment interaction, the temperature of the environment, and the timescales involved. The complexity of the interactions makes the detailed analysis of decoherence and quantum dissipation a complex task. In the next section, we will explore the important implications of these phenomena.

Decoherence and Dissipation: Key Implications

The study of decoherence and quantum dissipation has significant implications across various fields. Understanding and mitigating these phenomena are crucial for the development of robust quantum information processing technologies. Efforts to engineer quantum error correction codes and fault-tolerant quantum computing architectures aim to counteract the detrimental effects of decoherence and quantum dissipation, enabling the implementation of reliable quantum algorithms.

Decoherence and quantum dissipation also impact the sensitivity and precision of quantum measurements and sensing, sometimes even in positive ways. By understanding and controlling these processes, researchers can enhance the accuracy and resolution of quantum sensors and measurement devices. This has applications in fields such as precision metrology, gravitational wave detection, and quantum-enhanced imaging. Moreover, decoherence and quantum dissipation offer insights into the fundamental questions about the quantum-classical boundary and the emergence of classical behaviour from the quantum realm. Studying these phenomena helps elucidate the mechanisms by which the peculiarities of the quantum world give way to the familiar and predictable behaviour of the classical world. Decoherence and quantum dissipation are fundamental phenomena that bridge the gap between the quantum and classical realms and help us better understand the mysteries of the quantum world.

Transition from Quantum to Classical Behaviour

Recent studies have demonstrated that during the early inflationary period of the universe—when it expanded rapidly shortly after the Big Bang—quantum fluctuations evolved into classical inhomogeneities due to the universe's expansion and decoherence effects. This transition preserves the inflationary predictions for the current universe while making quantum properties indistinguishable from classical stochastic variations. This highlights the significance of the quantum-to-classical transition in understanding the structure of the universe.

Understanding the transition from quantum to classical behaviour is essential for comprehending the boundary that separates the microscopic and macroscopic realms. While the laws of classical physics govern our everyday experiences, the quantum world reveals strange phenomena, such as superposition and entanglement. This section will explore the fascinating process by which the peculiarities of the quantum realm give way to the familiar behaviour of the classical world.

Classical physics assumes an objective reality where object properties have definite values independent of observation. This transition implies that macroscopic objects exhibit measurable properties free from quantum uncertainties. Classical physics emerges as an approximation of quantum mechanics when particle numbers and interaction complexities exceed certain thresholds. Classical laws, such as Newton's laws of motion, arise as effective descriptions of the behaviour of macroscopic objects. This transition to classical behaviour highlights the role of measurement and observation in collapsing quantum wave functions. And when a quantum system interacts with its environment, it effectively undergoes a continuous process of measurement, leading to the collapse of its superposition states into classical-like states.

Moreover, the concept of decoherence is central to understanding quantum-to-classical transition. As mentioned earlier, when a quantum system becomes entangled with its surrounding environment, its delicate superposition states rapidly lose their coherence due to the spreading of information and the destruction of interference patterns. In this case, the environment acts as a measuring apparatus, interacting with the quantum system and effectively 'measuring' its properties. These interactions cause the quantum system to rapidly interact with a vast number of environmental particles, leading to the emergence of a classical-like behaviour. Therefore, we can see how process of decoherence clarifies why macroscopic objects, composed of an enormous number of particles, display classical behaviour instead of the bizarre and counterintuitive behaviour of the quantum world. And as the number of particles involved increases, the effects of decoherence become more pronounced, effectively erasing the quantum features and giving rise to the classical behaviour we observe in our daily lives. However, it is important to note that decoherence does not completely eliminate quantum effects. Even in macroscopic systems, traces of quantum behaviour can still be observed, but they become increasingly difficult to detect as the system interacts more strongly with its environment.

Overall, the transition from quantum to classical behaviour raises questions about the fundamental boundaries between the microscopic and macroscopic worlds. Understanding these boundaries is crucial for developing a comprehensive theory that unifies quantum mechanics and classical physics. The study of quantum-to-classical transition is an active area of research, with ongoing efforts to explore the limits of quantum coherence, understand the mechanisms of decoherence, and investigate novel approaches to control and manipulate quantum systems to maintain their coherence.

Impact of Environmental Interactions on Quantum Systems

Quantum systems, with their peculiarities of superposition and entanglement, are known for their delicate nature. However, these systems do not exist in isolation. They are constantly interacting with their surrounding environment, which can have a profound impact on their behaviour. Let's look at the intricate relationship between quantum systems and their environment, shedding light on the fascinating phenomenon of decoherence and its consequences.

The phenomenon of decoherence can be seen as the spreading of information from the quantum system to the environment, resulting in the destruction of interference patterns and the emergence of classical-like behaviour. In such a case, environmental interactions—the primary cause of decoherence—can manifest in various forms, such as collisions with particles, electromagnetic interactions, or fluctuations in temperature. These interactions can disrupt phase coherence, which is essential for the unique properties of quantum systems. This disruption leads to the suppression of interference effects and the rapid decay of quantum superpositions. Coherence is essential for maintaining quantum effects like superposition and entanglement, which are critical for quantum computations and other quantum phenomena. However, the loss of coherence due to environmental interactions limits the practicality of maintaining quantum behaviour in large-scale systems.

Thus, environmental interactions make it challenging to observe and manipulate quantum effects. The presence of the environment introduces noise and disturbances that can obscure quantum phenomena. This poses a challenge for researchers attempting to study and control quantum systems, as it becomes increasingly difficult to maintain their coherence and extract accurate information from them.

The impact of environmental interactions on quantum systems has spurred the development of quantum error correction techniques. These techniques aim to mitigate the effects of decoherence by encoding and protecting quantum information against environmental disturbances. By using redundancy and error-correcting codes, quantum systems can be shielded from the detrimental impact of environmental interactions, leading to more robust and reliable quantum computations and communication. However, there is an intriguing aspect to consider. In some cases, interactions with the environment can be beneficial. Recently, a model for quantum batteries was proposed where the environment, modelled using quantum Brownian motion, acts as a 'charger' for the system. This is possible due to the system's non-Markovian evolution. Non-Markovianity describes quantum systems where future dynamics depend not only on the current state but also on its past history. Unlike Markovian systems, which evolve based solely on their present state without memory of past interactions, non-Markovian systems incorporate memory effects, allowing the environment to positively influence the system's energy storage.

Understanding and controlling the impact of environmental interactions are essential for advancing quantum technologies. Efforts to design and engineer quantum devices, such as quantum computers and quantum sensors, require strategies to mitigate decoherence and minimise the effects of environmental noise. By effectively managing environmental interactions, researchers can enhance the stability and coherence of quantum systems, enabling more practical and efficient quantum technologies.

The study of environmental interactions with quantum systems raises fundamental questions about the nature of measurement, the emergence of classical behaviour, and the boundaries between the quantum and classical worlds. Exploring the impact of the environment on quantum systems provides insights into the foundational aspects of quantum mechanics and our understanding of reality. As we continue to explore and control the impact of the environment on quantum systems, we move closer to harnessing the potential of quantum technologies and expanding our understanding of the quantum world.

Self-Selected Fluctuations

Physicalist heuristics are inherently self-selecting in their fluctuations, a thesis I proposed, supported by Prof. Brian Josephson in 2020. Self-selected fluctuations in physics refer to the natural variability within entities such as quantum fields, which can be understood through both classical and quantum fluctuation concepts. In quantum systems, these fluctuations manifest as temporary changes in energy levels at specific points in space, as described by the Heisenberg uncertainty principle. Quantum fluctuations lead to the spontaneous creation and annihilation of virtual particles in a vacuum, which, while not directly observable, have measurable effects on physical phenomena. Notable examples include the Lamb shift in hydrogen and the Casimir effect, both of which arise from these fluctuations.

The Lamb shift demonstrates how vacuum fluctuations can affect the energy levels of electrons in atoms, while the Casimir effect shows the attractive force between closely spaced conductive plates due to changes in vacuum energy. The implications of self-selected fluctuations extend into quantum field theory, where they are crucial for understanding particle dynamics and interactions. In this framework, field fluctuations are essential for explaining phenomena such as the running of coupling constants and the renormalisation of particle masses. These fluctuations contribute to the effective mass and charge of elementary particles, which are vital for accurately predicting outcomes in particle physics experiments. Far from being random noise, these fluctuations are self-regulating phenomena arising from the field's underlying structure. By modelling and analysing these fluctuations, physicists gain valuable insights into the behaviour of fundamental forces and particles, deepening our understanding of the universe's structure.

Furthermore, the concept of self-selected fluctuations is closely tied to the renormalisation process in quantum field theory (QFT), where infinities in calculations are systematically managed. Renormalisation adjusts parameters in the theory to account for these fluctuations, effectively selecting a stable ground state from the chaotic behaviour of quantum fields. This process underscores how fluctuations can lead to emergent properties that define particle and field behaviour at various scales. For example, the running of coupling constants—how interaction strengths vary with energy scales—is a direct result of these self-selected fluctuations within quantum fields.

Self-selected fluctuations also significantly impact the formulation of effective field theories, which simplify high-energy physics into more manageable models. These theories rely on averaging out high-energy fluctuations at low energies, allowing for a clearer description of particle interactions.

This averaging process, a form of self-selection, identifies the relevant degrees of freedom that govern the system's dynamics at lower energies, enabling physicists to focus on the most impactful fluctuations without being overwhelmed by the full complexity of QFT.

In cosmology, self-selected fluctuations are crucial for explaining dark energy and the universe's expansion. The energy density associated with vacuum fluctuations is linked to the cosmological constant, which is key to understanding the accelerated expansion observed in distant galaxies. These fluctuations impact not only local particle interactions but also the large-scale structure of the cosmos. The connection between vacuum fluctuations and cosmological phenomena highlights their role in shaping the universe's evolution from the Big Bang to the present.

During the universe's early moments, quantum field fluctuations are believed to have seeded the large-scale structure we see today. These initial fluctuations, amplified during inflation, led to the formation of galaxies and clusters, illustrating the profound impact of self-selected fluctuations on cosmic evolution. Additionally, research on quantum metric fluctuations has explored their effects on cosmological evolution. These fluctuations can alter the gravitational field equations that govern the universe's dynamics, offering new insights into cosmic structures. By analysing their influence on Friedmann-Lemaître-Robertson-Walker geometries, researchers can deepen their understanding of the universe's expansion and the role of dark energy.

Self-selected fluctuations also have profound implications in the study of complex systems, such as turbulence and many-body systems. In these contexts, fluctuations can lead to the emergence of patterns and instabilities that defy classical predictions. For example, in fluid dynamics, quantum fluctuations can influence turbulent flow behaviours, while in condensed matter physics, they may drive phase transitions and critical phenomena. Understanding these fluctuations is essential for developing accurate models that describe the dynamics and behaviours of such systems.

Moreover, advancements in experimental techniques have enabled researchers to directly measure and manipulate vacuum fluctuations, leading to innovative applications in quantum technology. For instance, the dynamical Casimir effect demonstrates how mechanical motion can excite vacuum fields, potentially transferring energy from the vacuum to macroscopic systems. These insights pave the way for new applications in quantum computing, precision measurements, and other emerging technologies. The exploration of self-selected fluctuations remains a vibrant area of research, with the potential to enhance our understanding of fundamental physics and its practical applications.

Additionally, machine learning has emerged as a crucial tool for studying quantum dissipative dynamics, especially when traditional methods struggle with the computational complexity of modelling open quantum systems. Classical approaches, whether fully classical, quantum-classical, or fully quantum, often face high computational costs, limiting their applicability. Recent advancements in machine learning have addressed these challenges by enabling more efficient predictions of quantum dynamics. This includes tasks such as predicting molecular configurations, relaxation dynamics, and excitation transfer. The open-source software package MLQD supports these studies by providing tools for kernel ridge regression, convolutional neural networks, hyperparameter optimisation, and result visualisation. Its integration with cloud computing platforms further enhances performance, making it a valuable resource for advancing research in quantum dynamics.

The Knill-Laflamme conditions are essential for identifying exact quantum error correction codes, as they ensure that the inner product of any two distinct quantum codewords remains invariant under error operators, thus preserving encoded information despite errors. However, these conditions are highly restrictive and may not always lead to the most effective codes. To overcome this limitation, a generalised performance metric known as near-optimal channel fidelity has been introduced. This metric provides a precise, optimisation-free evaluation of code performance across different codes and noise conditions, offering a tight two-sided bound on optimal performance.

Recent research has highlighted the significant numerical advantages of this metric, enabling simulations of larger systems, such as oscillators with hundreds of excitations, which were previously impractical. Notably, the Gottesman-Kitaev-Preskill code demonstrates improved performance with increasing energy, achieving an asymptotic limit at infinite energy, in contrast to other oscillator codes.

If we consider the 'prehistory' of the concept of decoherence, we can find early hints in the writings of Werner Heisenberg, who alluded to the role of the environment. However, he did not invoke the notion of 'spooky action at a distance'. For example, Heisenberg suggested that:

> 'The interference terms are . . . removed by the partly undefined interactions of the measuring apparatus, with the system and with the rest of the world.'

Currently, research is focused on understanding the mechanics and non-equilibrium thermodynamics of decoherence, under the assumption that the system's design (Hamiltonian) remains constant and that there is no direct energy exchange between the system and its surroundings. For instance, in the case of a qubit within a degenerate Fermi gas, it has been observed that the heat generated by the system follows an integral fluctuation relation and is related to the production of entropy associated with the system's energy state.

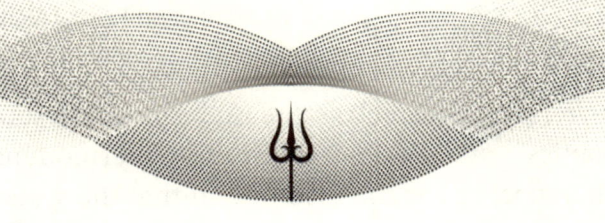

Chapter 9
Quantum Erasers and Decoherence

In elocution 101, we are taught to be coherent, but achieving coherence is significantly more challenging in the quantum realm compared to everyday life. Decoherence is central to our understanding of quantum mechanics, addressing fundamental questions about why quantum systems lose their 'quantum-ness'. Is this loss an inherent trait of quantum systems, or is it the result of interactions with their surroundings? Decoherence helps to explore and answer these crucial questions.

In this chapter, we will delve deeper into the details of decoherence, explore the concept of the quantum eraser, and discuss their roles in advancing quantum technology.

Decoherence in Quantum Systems

Quantum decoherence is a fundamental aspect of quantum mechanics, describing the process by which quantum systems lose their coherence and begin to exhibit classical-like behaviour. While theoretical frameworks and mathematical models provide insights into the mechanisms of decoherence, experimental observations and verifications are crucial for understanding and confirming its existence.

Experimental investigations of quantum decoherence involve probing the behaviour of quantum systems and observing their evolution over time. By subjecting quantum systems to controlled interactions with their environment, researchers aim to elucidate the processes that cause the loss of coherence and the emergence of classical behaviour. These experiments provide invaluable insights into the limitations and challenges faced in maintaining and manipulating quantum states.

Several experimental techniques have been developed to observe and verify quantum decoherence. These experiments involve the manipulation of quantum systems to create superposition states and then subject them to environmental interactions. By observing interference patterns, researchers can assess the coherence of the system and investigate the effects of the environment. These deviations are particularly evident in single-photon interference experiments, which provide a unique platform to study this phenomenon. By using single photons and carefully designed setups, researchers can explore the impact of environmental interactions on the interference patterns. For instance, double-slit experiments with single photons reveal the loss of interference when the photons interact with their environment, leading to a reduction in coherence.

Additionally, quantum optics and cavity quantum electrodynamics (QED) experiments allow for the manipulation and control of light-matter interactions at the quantum level. These setups involve trapping atoms or ions in cavities, creating controlled environments to investigate decoherence. One effective platform for studying decoherence is spin systems, such as those found in nuclear magnetic resonance (NMR) experiments. These experiments involve the manipulation of spins in atomic nuclei using magnetic fields. This understanding is further enhanced by experiments in quantum computing and quantum information processing, which highlight the challenges posed by decoherence. Researchers utilise qubits, the building blocks of quantum computers, to implement quantum algorithms and perform calculations. Monitoring the behaviour of qubits under environmental interactions allows researchers to assess the impact of decoherence on the stability and reliability of quantum computations. These experimental observations not only confirm the existence of decoherence but also provide valuable information about its characteristics and timescales. This information has practical implications and offers valuable guidance for the design and development of quantum technologies, including quantum computers, communication systems, and sensing devices.

Quantum Eraser Experiments and Empiricism of Decoherence

Quantum eraser experiments have long been a subject of fascination and debate within the realm of quantum mechanics. These experiments challenge our understanding of the nature of reality and the role of observation in shaping quantum phenomena. In this section, we delve into the captivating world of quantum eraser experiments and explore their implications for the empirical metaphysics of decoherence.

At the core of quantum eraser experiments is the principle of wave-particle duality, which asserts that quantum particles can exhibit both wave-like and particle-like behaviour. These experiments seek to investigate the interplay between interference phenomena and decoherence and typically involve the use of entangled particles or entangled quantum systems, such as pairs of entangled photons. In these experiments, entangled particles are split into two separate paths. Upon recombining the particles, interference patterns emerge and are observed. However, what makes quantum eraser experiments particularly intriguing is the possibility of 'erasing' or restoring the interference pattern after certain measurements or observations have been performed. This is achieved through additional measurements and post-selection procedures, where the erased interference patterns reemerge.

In summary, quantum eraser experiments continue to provide deep insights into the mysterious nature of quantum reality and the interplay between observation and the behaviour of quantum systems.

Decoherence and Empirical Metaphysics

The empirical metaphysics of decoherence comes into play when we consider the implications of these experiments for our understanding of the nature of reality. Empirical metaphysics emphasises the importance of empirical evidence and observation in gaining knowledge about the world. Meaning, it asserts that the nature of reality is revealed through observation and measurement.

In the context of quantum eraser experiments, empirical metaphysics raises questions about the role of observation and its impact on the behaviour of quantum systems. One interpretation suggests that the act of observation or measurement influences the state of the quantum system, determining whether interference patterns are observed or not. From this perspective, we can conclude that observation plays a direct role in shaping the behaviour of quantum systems. However, another interpretation arises from the perspective of decoherence. According to this view, the restoration of interference patterns in quantum eraser experiments can be seen as a consequence of controlling and manipulating the environmental interactions to counteract the effects of decoherence. This interpretation challenges the notion of the direct role of observation in influencing the behaviour of the quantum system.

The empirical metaphysics of decoherence invites a deeper exploration of the intricate relationship between observation, environmental interactions, and the nature of quantum systems. It highlights the complexities of understanding the boundaries between the quantum and classical realms and the role of observation in revealing the true nature of reality. Quantum eraser experiments offer a captivating glimpse into the interplay between interference phenomena and decoherence. They raise important questions about the empirical metaphysics of decoherence and the role of observation in shaping quantum behaviour. By examining the mechanisms and interpretations of these experiments, we deepen our understanding of the delicate nature of quantum systems and the complex relationship between observation, environmental interactions, and the empirical metaphysics of decoherence.

Challenges Posed by Decoherence to Quantum Technology

Quantum technology is rapidly emerging as a revolutionary field with the potential to transform various aspects of our lives, from computing and communication to sensing and cryptography. It is based on the principles of quantum mechanics, which govern the behaviour of particles at the atomic and subatomic levels. However, one of the significant hurdles that quantum technology faces is the phenomenon of decoherence. To comprehend the challenges posed by decoherence, it is crucial to understand the phenomenon itself.

As discussed earlier, in a quantum system, such as a qubit (the basic unit of quantum information), coherence refers to the superposition of multiple states. These states can exist simultaneously and undergo complex computations. However, when a quantum system interacts with its environment, the delicate superposition is disrupted, leading to decoherence. The system loses its quantum properties and behaves classically, rendering it useless for quantum technology applications. Several factors contribute to the decoherence of quantum systems. One of the primary culprits is environmental noise or interaction with the surrounding environment. Factors such as temperature fluctuations, electromagnetic radiation, and material impurities can introduce noise and disturb the coherence of quantum systems. Additionally, manufacturing imperfections, such as defects in qubit design or fabrication, can also lead to decoherence. These imperfections create unintended interactions that disrupt the desired quantum behaviour.

Decoherence poses a significant challenge to quantum computing, one of the most promising applications of quantum technology. Quantum computers leverage the ability of qubits to exist in superposition states to perform computations at an unprecedented speed. However, as the number of qubits and the complexity of computations increase, so does the susceptibility to decoherence. The fragility of quantum states makes it challenging to maintain the coherence required for accurate and reliable calculations.

To mitigate the impact of decoherence on quantum computing, researchers are actively exploring techniques such as error correction and fault-tolerant quantum computation. Error correction algorithms and protocols aim to detect and correct errors introduced by decoherence and other sources of noise. By redundantly encoding quantum information, errors can be identified and rectified, preserving the coherence of the quantum system. These codes use multiple qubits to encode a single logical qubit, allowing for error detection and correction. Examples of quantum error correction codes include the surface code, the topological colour code, and the stabiliser codes. These codes provide a framework for protecting quantum information from the detrimental effects of decoherence, but they also require significant resources and specialised hardware. However, implementing error correction in practical quantum computers remains a formidable challenge due to the additional overhead it introduces.

Decoherence-free subspaces (DFS) offer an alternative approach to tackle decoherence. DFS involves designing a quantum system so that certain states remain immune to decoherence while preserving the desired quantum properties. By carefully selecting and manipulating the system's states, it is possible to create protected subspaces where quantum information can be stored and processed without significant decoherence. However, implementing DFS requires precise control over the system and is limited to specific applications.

Another strategy to combat decoherence is quantum error avoidance, which focuses on designing quantum algorithms and hardware architectures that minimise the susceptibility to errors. By carefully crafting quantum gates and optimising the quantum circuit layout, it is possible to reduce the impact of decoherence. Additionally, advances in qubit fabrication and materials science can contribute to creating more robust quantum systems less prone to decoherence. Moreover, real-time error monitoring allows for rapid detection and immediate correction, maintaining the quantum system in a coherent state. These techniques often rely on feedback control and adaptive algorithms to counteract the disruptive effects of decoherence. However, the implementation of such techniques requires high-speed measurement and control capabilities.

To summarise, while decoherence poses significant challenges to the development and practical implementation of quantum technology, researchers and scientists are actively exploring various strategies to overcome these challenges. Techniques such as error correction, fault tolerance, decoherence-free subspaces, error avoidance, and error suppression offer promising avenues for mitigating the effects of decoherence. As our understanding of decoherence improves and technological advancements continue, we can hope to unlock the full potential of quantum technology and realise its transformative impact on multiple domains.

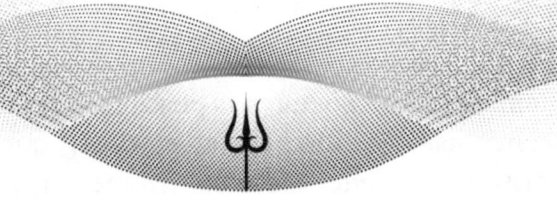

Chapter 10
The Plinth of Entangled Realities

*E*verything is interconnected, and this idea extends beyond philosophy into the realm of quantum mechanics. Entanglement, a phenomenon where particles become interconnected in such a way that the state of one instantaneously influences the state of another, is one of the most intriguing aspects of quantum physics, drawing interest from both scientists and philosophers. While its existence has been experimentally confirmed and it plays a critical role in various quantum technologies, entanglement also raises profound questions that challenge our understanding of reality. In this chapter, we will explore these fundamental questions about entanglement and examine their implications for our comprehension of the quantum world.

Entanglement: Challenging Classical Concepts

One of the crucial aspects of entanglement is the concept of non-locality, which means that in entangled systems, measurements made on one particle instantaneously affect the state of another, regardless of the distance between them. Another fundamental issue related to entanglement is the measurement problem and the concept of wave function collapse. This phenomenon suggests that the act of observation or measurement can alter the state of a quantum system, challenging the classical notion of an objective reality independent of observation.

Entanglement also brings to light the idea of quantum non-separability, where the states of entangled particles cannot be described independently. This non-separability complicates our understanding of information flow and the conservation of information in entangled systems. Additionally, the existence of entangled systems raises the intriguing question of whether information can ever be truly destroyed, as entanglement appears to suggest a non-local correlation that goes beyond classical concepts of information transfer.

Another concept related to entanglement is the arrow of time, which describes the asymmetry between cause and effect in our everyday experience. Entanglement is time-symmetric, meaning that entangled particles can exist in a superposition of states that span the past, present, and future. However, when measurements are made, the collapse of the wave function introduces a directionality that aligns with the arrow of time. This raises the question of how entanglement fits within our understanding of time and whether it contributes to the emergence of the arrow of time.

Entanglement thus stands as a fundamental pillar of quantum mechanics, providing deep insights into the behaviour of quantum systems. Yet, it also raises profound questions about the nature of reality, the role of observation, the measurement process, and the flow of information. As researchers continue to explore these mysteries, we edge closer to uncovering the fundamental nature of the quantum realm and its implications for our understanding of the universe.

Insights from Quantum Entanglement

Entanglement, a phenomenon in quantum mechanics, has profound implications not only for our understanding of the nature of reality at a fundamental level but also for our understanding of our individual existence. Let's closely examine the most significant empirical, philosophical, and metaphysical implications of entanglement in detail.

1. **Consciousness and Interconnectedness**

   Entanglement challenges our understanding of individual existence by suggesting an interconnectedness that extends beyond the classical boundaries of space and time. This phenomenon implies that we are intricately connected to the broader fabric of reality, with our actions and experiences potentially having ripple effects beyond our immediate perception. It invites us to reconsider the idea of an isolated, separate self and encourages a shift towards a more holistic view of our place in the universe. This understanding of interconnectedness raises profound questions about the nature of consciousness and its role in shaping our subjective experience of reality.

2. **Observer and Consciousness**

   The role of the observer and consciousness in quantum mechanics, particularly in relation to entanglement, remains a subject of intense debate. Some interpretations suggest that the act of observation or measurement can influence the state of a quantum system, leading to the collapse of the wave function. While the exact mechanism of this influence is still debated, with some attributing it to decoherence alone, others consider the possibility of consciousness playing a role. This connection between observation and quantum phenomena raises intriguing questions about the nature of consciousness and its potential influence on physical reality.

3. Wave-Particle Duality and Consciousness

Recent research has challenged traditional notions of wave-particle duality, suggesting that global multi-partite entanglement may impose restrictions on this fundamental concept. New insights propose viewing wave-particle duality through the lens of information conservation in multi-path interferometers, with uncertainty as a core theme. A newly formulated complementarity relation links wave characteristics, particle characteristics, and the mixedness of quantum states, offering a fresh perspective on traditional approaches. This evolving understanding of wave-particle duality and its relationship to entanglement may have implications for our conception of consciousness and its role in quantum phenomena.

4. Consciousness and Free Will

Entanglement raises profound questions about the concepts of free will and causality. The instantaneous and probabilistic correlation of entangled particles across vast distances challenges the notion of a deterministic universe governed by simple cause and effect. This suggests the possibility of non-local influences that transcend our classical understanding of causality, inviting us to question the traditional dichotomy between determinism and free will. If we were to build a philosophy based on the metaphysics of quantum mechanics, it suggests a more intricate interplay of factors influencing our choices and actions. This perspective could potentially transform our understanding of consciousness and its role in decision-making, offering a deeper insight into how our perceptions and decisions are interconnected with the quantum realm.

5. Multiverse and Consciousness

Entanglement is closely tied to the concept of the multiverse, where multiple branches of reality coexist simultaneously, as per the Many Worlds Interpretation of quantum mechanics. According to this hypothesis, the measurement of quantum particles can create multiple parallel universes or branches, each representing a different outcome. Entanglement plays a critical role in facilitating this measurement process, particularly in decoherence theory. This perspective invites us to consider the expansive nature of our personal existence and the richness of potential futures that lie before us, potentially reshaping our understanding of consciousness and its relationship to multiple realities.

6. Perception and Reality Construction

Entanglement challenges our understanding of the relationship between perception and the construction of reality. The interconnectedness suggested by entanglement implies that our perception and observation of the world may actively shape the reality we experience. This challenges the notion of a fixed, objective reality and highlights the dynamic nature of our individual experiences. It encourages us to explore how our subjective perception, influenced by our consciousness and beliefs, interacts with the external world to shape our personal reality, potentially blurring the lines between observer and observed.

In summary, it is clear that quantum entanglement not only challenges our classical understanding but also opens the door to new possibilities in science and technology. Its exploration has paved the way for future research and innovations that could redefine our approach to information, communication, and even the nature of reality itself. Embracing the complexities of entanglement invites us to continually question and expand the boundaries of our knowledge, driving progress in both theoretical and applied physics.

Section III

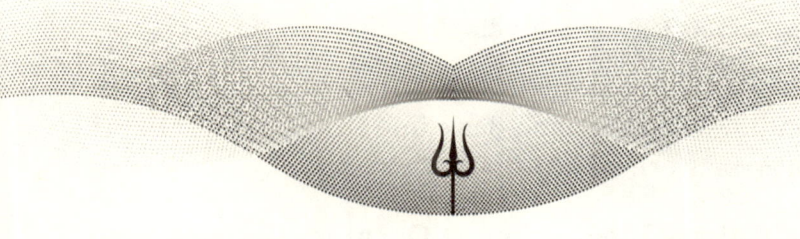

Chapter 11

Quantrika: Beyond the Veil

The intriguing relationship between modern quantum mechanics and ancient philosophical traditions has long fascinated scholars. Early on, researchers like Huston Smith recognised the parallels between the quantum realm and certain Eastern traditions. Smith, for instance, highlighted how quantum mechanics interestingly correlates with the metaphysics of the Tiwi, an Aboriginal people of Melville and Bathurst Islands in northern Australia.

> 'Ancient wisdom and quantum physicists make unlikely bedfellows: In quantum mechanics, the observer determines (or even brings into being) what is observed, and so, too, for the Tiwis, who dissolve the distinction between themselves and the cosmos. In quantum physics, subatomic particles influence each other from a distance, and this tallies with the aboriginal view, in which people, animals, rocks, and trees all weave together in the same interwoven fabric.'

In this chapter, we will closely examine the intriguing connections and parallels between Trika Shaivism and quantum mechanics and delve into their potential to bridge the gap between spirituality and physics.

Foundational Intersections of Quantum Mechanics and Trika Shaiva Philosophy

As we have seen so far, both Trika Shaivism and quantum mechanics explore the nature of reality, consciousness, and the interplay between the observer and the observed. And while they may appear to be divergent in their approaches, both traditions provide valuable insights into the essence of the individual and its relationship to the broader fabric of reality.

Let's explore some of the intriguing connections between these two fields and see how they enhance our understanding of consciousness and existence.

1. Trika Shaivism, at its core, emphasises non-dualism, asserting that all of existence is an interconnected, unified whole. Similarly, quantum mechanics reveals that at the fundamental level of reality, particles and phenomena are not separate entities but interconnected aspects of a unified quantum system. The concept of entanglement in quantum mechanics reflects the interconnectedness and non-dual nature of reality, aligning with the non-dualistic teachings of Trika Shaivism.

2. Quantum mechanics introduces the role of the observer and highlights the fundamental influence of observation on the behaviour of quantum systems. Trika Shaivism reflects a similar notion where the practitioner seeks to transcend the dualistic

perspective of being separate from the observed world and recognises the interplay between the subject and object. Both traditions emphasise the significance of the observer in shaping and co-creating reality.

3. The wave-particle duality suggests that particles can exhibit both wave-like and particle-like behaviour depending on the context of measurement. Similarly, in Trika Shaivism, reality is seen as a dance of manifestation and dissolution, where the underlying essence can take on different forms. The concept of Shakti, the dynamic creative power in Trika Shaivism, aligns with the notion of wave-like potentiality in quantum mechanics.

   Moreover, the concept of dynamism inherent in Shakti, the dynamic creative power in Trika Shaivism, and the concept of spanda, or the vibrational pulse of consciousness, reflect the dynamic and ever-changing nature of reality. This is similar to how particles and waves exhibit different behaviours based on the context of observation in quantum mechanics. Just as wave-particle duality indicates that particles can exhibit both wave-like and particle-like properties depending on how they are observed, the Trika perspective acknowledges that the same underlying essence can manifest in many different forms.

4. Trika Shaivism explores the concept of universal consciousness, recognising the inherent unity and interconnectedness of all beings. Quantum mechanics, on the other hand, presents the idea of a unified field of quantum-ness, a fundamental quantum state that

underlies all particles and phenomena. The concept of entanglement further supports this notion of universal interconnectedness.

5. Trika Shaivism emphasises practices such as meditation and self-inquiry to transcend the limited perception of the ego and awaken the expansive nature of consciousness. Similarly, quantum mechanics reveals the limitations of classical determinism and invites us to transcend our classical intuitions to grasp the subtle interconnectedness and potentiality inherent in quantum systems.

The foundational intersection of Trika Shaivism and quantum mechanics offers a fertile ground for exploration, bridging the ancient wisdom of Kashmir Shaivism with the modern insights of quantum physics. The shared concepts of non-dualism, the role of the observer, manifestation, unity of consciousness, and the expansion of awareness highlight the deep connections between these seemingly disparate domains. Moreover, this confluence opens doors for a deeper understanding of the nature of reality, consciousness, and our place in the interconnected web of existence.

The Individual in Trika Shaivism and Quantum Mechanics

In Trika Shaivism, the individual is viewed as an expression of the ultimate reality, referred to as 'Shiva consciousness'. The tradition teaches that the individual, known as the jiva, is not separate from the divine, but rather an extension of it. It further elaborates that the true essence of the individual is non-different from the ultimate reality, transcending the limitations of the ego and the illusory sense of separate identity. Similarly, quantum mechanics reveals that at the fundamental level, particles and phenomena are interconnected aspects of a unified quantum system. This non-duality challenges the notion of a fixed and separate individual identity, inviting us to question the boundaries of individuation and embrace the interconnectedness that underlies existence. Furthermore, entanglement—a central concept in quantum mechanics—has profound implications for our understanding of individuality. It suggests that quantum particles can become intrinsically correlated, regardless of spatial separation. From the perspective of entanglement, individuals are intimately interconnected, sharing a subtle connection that transcends classical notions of space and time. This interconnectedness challenges the isolated concept of the individual and invites us to consider the collective aspect of existence. It emphasises the interdependence and shared nature of our individual experiences.

Trika Shaivism also states that the jiva possesses unique non-absolute qualities and experiences that contribute to the diversity of existence. Similarly, quantum mechanics reveals that the wave-particle duality allows for the manifestation of different aspects of reality depending on the context of measurement. This intersection suggests that individuals are both unique expressions of the ultimate reality and active participants in the process of reality manifestation. It is important to recognise that this connection is a resonance that may bridge the ontology-epistemology divide.

Lastly, both traditions emphasise the importance of expanding consciousness and self-realisation. Trika Shaivism encourages the individual to transcend egoic limitation and awaken to their true nature as Shiva consciousness.

This process involves expanding awareness, purifying the mind, and experiencing union with the divine. In quantum mechanics, the elusive consciousness and the observer effect prompt us to expand our understanding and embrace the inherent potentiality of existence. In this view, the individual is seen as an agent of transformation, capable of expanding consciousness and participating in the co-creation of reality. However, it remains to be seen whether our conceptualisations of consciousness, such as Tononi's Integrated Information Theory, will converge with ideas of decoherence and measurement.

Overall, the understanding of the individual in Trika Shaivism and quantum mechanics is a complex and multifaceted topic. While Trika Shaivism emphasises the divinity of the individual and the potential for self-realisation, quantum mechanics highlights the active role of the observer in shaping reality. Both traditions recognise the interconnectedness and non-dual nature of existence, challenging the notion of separate individuality. Exploring the individual within these frameworks invites us to transcend limited perspectives, expand our understanding, and embrace the profound interconnectedness that underpins the fabric of reality.

Correlations and Agency in Trika Shaivism and Quantum Mechanics

Trika Shaivism and quantum mechanics originate from different cultural and intellectual backgrounds, yet there are intriguing parallels in their exploration of correlations and agency. This section explores how both traditions examine these themes, highlighting the shared insights and contrasting perspectives that emerge.

1. Correlations

Trika Shaivism views reality as a divine manifestation where the ultimate reality, often referred to as Shiva, expresses itself through diverse forms. This tradition recognises the interplay between Shiva, representing ultimate reality, and Shakti, the dynamic creative energy, in shaping reality. These divine correlations reflect the interconnectedness and interdependence of all phenomena. The individual participates in this divine dance and experiences correlations that reveal the underlying unity of existence. On the other hand, quantum mechanics examines correlations between quantum systems that defy classical notions of causality and determinism. Entanglement, for instance, demonstrates intrinsic correlations between quantum particles regardless of distance. Measurements on one entangled particle instantaneously influence the state of another, highlighting profound interconnectedness. Quantum mechanics acknowledges correlations as pivotal in shaping the behaviour and properties of quantum systems, underscoring the non-local nature of these relationships.

2. Agency

Trika Shaivism emphasises the concept of agency, viewing individual experience as the play of consciousness. While recognising the ultimate reality as the supreme agent, Trika Shaivism also acknowledges that the individual possesses a limited sense of agency. It further states that through practices such as meditation, self-inquiry, and devotion, individuals can cultivate awareness and align their limited agency with the divine will. Moreover, it also highlights the transformative potential of agency, allowing individuals to participate consciously in the divine dance and align their actions with the harmonious unfolding of reality. On this topic, quantum mechanics introduces the role of the observer and its influence on the observed quantum system. The act of observation collapses the wave function, thereby determining the outcome of measurements. This recognition of the observer's agency challenges classical notions of objectivity and determinism. Altogether, quantum mechanics reveals that the observer plays an active role in shaping the observed reality, highlighting the role of agency in the quantum realm.

In conclusion, Trika Shaivism and quantum mechanics both emphasise the interplay between correlations and agency. Exploring these shared insights and contrasting perspectives enriches our understanding of how correlations and agency intersect within the complex fabric of existence.

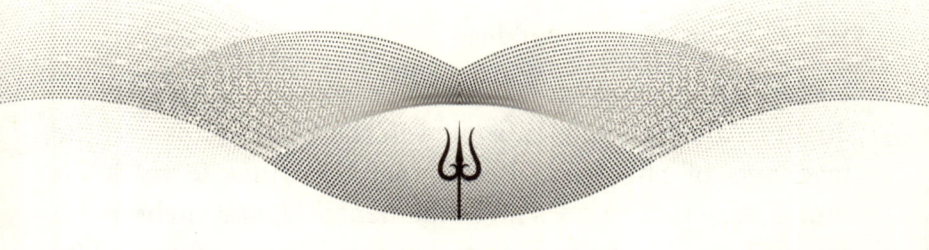

Chapter 12
Non-duality and Complementarity

Non-duality is the understanding that beneath the diversity and multiplicity of experiences lies a single, infinite, and indivisible reality. Complementarity, on the other hand, is the principle that a comprehensive description of a phenomenon, especially in the quantum realm, requires two distinct and mutually exclusive perspectives—such as the wave and particle descriptions of light—which together offer a complete understanding of the system.

The concepts of non-duality and complementarity have been explored extensively by philosophical and spiritual traditions worldwide. Both Eastern and Western perspectives provide unique insights into the nature of reality, the interplay of opposites, and the quest for unity. This chapter aims to explore ancient and modern perspectives from both traditions, examining their shared insights and contrasting viewpoints. By doing so, we aim to deepen our understanding of these profound concepts.

Oriental and Occidental Perspectives on Non-duality and Complementarity

Oriental philosophies, such as Advaita Vedanta, Zen Buddhism, and Taoism, emphasise the principle of non-duality. Advaita Vedanta, rooted in the Upanishads and the teachings of Shankara, asserts that the ultimate reality is non-dual, where there is no fundamental separation between the individual and the divine. Zen Buddhism, influenced by Mahayana Buddhism, explores the concept of non-duality through practices such as meditation and *koan* introspection. Taoism elucidates the unity and interconnectedness of all things. These Oriental traditions highlight the realisation of non-duality as a means to transcend the illusory sense of separation and to awaken to the ultimate reality.

On the other hand, Occidental philosophies, particularly in ancient Greece with figures like Heraclitus and Parmenides, contain elements of complementarity. Heraclitus proposed the notion of the 'unity of opposites,' where contrasting elements are interconnected and give rise to the dynamic flow of reality. While Parmenides stressed the unity and immutability of being, suggesting that the apparent diversity of the world is illusory. These ancient Greek philosophies hinted at the complementary nature of opposites and the underlying unity that transcends apparent distinctions.

In contemporary times, the concept of non-duality has found resonance in various philosophical and spiritual movements. The works of 19th-century philosophers such as Arthur Schopenhauer and Friedrich Nietzsche explored the concept of the 'will' and the dissolution of subject-object dualism. In the 20th century, the non-dualistic teachings of figures like Jiddu Krishnamurti and Ramana Maharshi gained global prominence. Krishnamurti emphasised the realisation of non-duality through self-inquiry and the transcendence of conditioned patterns of thought, whereas Ramana Maharshi advocated self-inquiry into the nature of the 'I' to realise the non-dual Self. These modern perspectives continue to inspire individuals to transcend the boundaries of duality and awaken to the underlying unity of existence.

In the realm of modern physics, complementarity stands as a foundational concept, particularly within quantum mechanics. Niels Bohr, one of the founders of quantum theory, proposed the principle of complementarity. According to this principle, particles and phenomena can exhibit both wave-like and particle-like characteristics depending on the context of measurement. Werner Heisenberg's uncertainty principle further highlights the complementary nature of certain physical quantities, such as position and momentum, emphasising the limits of simultaneous measurement. The concept of complementarity in quantum mechanics challenges classical notions of determinism and reveals the interplay of complementary aspects in the quantum realm.

While non-duality and complementarity may appear as distinct concepts, there is a potential for integrating and synthesising these perspectives. In the quest for a more comprehensive understanding of reality, the integration of these perspectives can enrich our exploration of the interconnected nature of existence. Additionally, the concepts of non-duality and complementarity can profoundly impact our personal and global perspectives. On a personal level, embracing non-duality can lead to a deeper sense of unity, compassion, and interconnectedness with all beings. Furthermore, recognising the complementarity of opposites allows us to appreciate the richness of diversity and embrace the inherent harmony in the interplay of contrasting elements. From a global perspective, integrating non-duality and complementarity can facilitate dialogue, understanding, and cooperation across diverse cultures and belief systems, fostering a more inclusive and holistic approach to global challenges.

In summary, both ancient and modern traditions emphasise realising non-duality as a way to transcend dualistic thinking and realise the underlying unity of existence. Exploring these perspectives, both individually and in combination, can transform our understanding of reality, expand our consciousness, and foster a more inclusive and harmonious worldview.

Beyond Duality: Trika Shaivism's Perspective on Reality

In Trika Shaivism, reality is perceived as non-dual, meaning there is no inherent separation between the individual and the divine. This perspective challenges the conventional notions of duality and invites a holistic understanding of existence. This section will explore Trika Shaivism's non-dualistic perspective on reality, examining its philosophical foundations, key concepts, and implications for our understanding of ultimate reality.

Dualistic thinking is characterised by the separation of subject and object, good and bad, and other dualities that permeate our perception. Trika Shaivism emphasises the transcendence of these dualistic categories and the recognition of the underlying unity that encompasses all apparent divisions. By transcending dualistic thinking, individuals can expand their consciousness and experience the inherent harmony and interconnectivity of all existence.

Trika Shaivism shares the non-dualistic ideology with Advaita Vedanta, a prominent non-dualistic philosophical tradition in India. Advaita Vedanta teaches that the ultimate reality, often referred to as brahman, is non-dual, where there is no fundamental separation between the individual self (atman) and the universal consciousness (brahman). Similarly, Trika Shaivism asserts that the individual, known as the jiva, is not separate from the divine, but an extension of it. This non-dualistic perspective emphasises the oneness of all existence where all individuals are seen as expressions of divine consciousness.

Trika Shaivism also recognises that the identification with the ego creates an illusion of individuality and duality, leading to a sense of separation. This illusion leads to a fragmented perception of reality and limits our understanding of our true nature. Trika Shaivism encourages individuals to transcend this illusion and recognise the underlying unity that transcends apparent divisions. By embracing non-duality and letting go of the identification with the ego, individuals can experience a profound shift in their perception of reality and their interconnectedness with all beings. Moreover, Trika Shaivism describes reality as a dance between Shiva and Shakti, the divine masculine and feminine principles that bring forth the manifestation of the universe. Shaivism teaches that recognising and aligning with this divine dance can deepen one's understanding of non-duality and facilitate the transcendence of limited perspectives.

The non-dualistic perspective of Trika Shaivism has profound implications for self-discovery and liberation. According to Trika Shaivism, realising non-duality leads to self-realisation, where individuals recognise their true nature as an expression of divine consciousness. This realisation liberates them from ego-based identifications. Trika Shaivism underscores the transformative power of non-duality to free individuals from dualistic constraints and awaken them to their existential truth.

In conclusion, Trika Shaivism's non-dualistic view offers deep insights into existence and our connection with the divine. By acknowledging the illusion of separation and embracing non-duality, individuals can transcend limited identities and discover the underlying unity that pervades all aspects of reality. This perspective not only enriches our understanding of reality but also cultivates harmony, freedom, and interconnectedness with all existence.

Quantum Insights: Complementarity and Wave-particle Duality

In the previous section, we discussed how Trika Shaivism posits the non-dual nature of existence. In this section, we will explore how two key concepts from quantum mechanics, complementarity and wave-particle duality, support this concept and reveal the profound interconnectedness of reality, highlighting the non-dual nature of existence. These concepts offer profound insights into the nature of quantum systems and highlight the inherent limitations of our observations. In this section, we delve into the principle of complementarity and wave-particle duality, exploring their significance, experimental validation, and implications for our comprehension of the quantum realm.

1. Wave-particle Duality

Wave-particle duality is a fundamental concept in quantum mechanics that highlights the dual nature of particles. It asserts that particles, such as electrons and photons, can manifest both wave-like and particle-like characteristics depending on the manner of observation or measurement. When particles undergo experiments focused on their position, they exhibit particle-like traits, such as localised position and definite momentum. Conversely, in experiments examining interference or diffraction patterns, particles display wave-like traits, such as interference fringes and diffraction patterns.

The wave-particle duality of quantum systems finds validation in numerous experimental observations. One pivotal experiment demonstrating this duality is the double-slit experiment. In this experiment, a beam of particles is directed at a barrier featuring two slits. When these particles are observed individually, they behave as discrete particles and create a pattern of two distinct bands on the screen behind the slits. However, when the particles pass through the slits without observation, they exhibit an interference pattern, indicative of their wave-like nature. The double-slit experiment underscores the wave-particle duality in quantum systems, where particles manifest both wave-like and particle-like characteristics contingent upon the mode of measurement or observation.

2. Complementarity

The principle of complementarity, introduced by Niels Bohr, posits that certain properties of quantum systems cannot be observed simultaneously with precision. In other words, it establishes a fundamental limit on the knowledge attainable about a quantum system. According to this principle, specific attributes of a quantum entity—like its position and momentum, or its wave-like and particle-like behaviours—are mutually exclusive. This implies that selecting one type of measurement precludes simultaneous precise measurement of its complementary attribute. Observing both aspects simultaneously would only be possible if certain constraints are applied to various degrees of freedom.

The principle of complementarity and wave-particle duality profoundly impacts our understanding of quantum reality. These concepts defy classical intuitions, which presume objects possess singular, well-defined states at any given time. Quantum mechanics, however, discloses that particles occupy superpositions of states until measured or observed, and their conduct is intrinsically probabilistic. This probabilistic nature signifies that we can forecast probabilities of outcomes rather than pinpoint specific predictions with certainty.

One intriguing aspect of the principles of complementarity and wave-particle duality is how they reinforce the primacy of the observer in the measurement process in quantum mechanics. Quantum mechanics suggests that the act of observation influences the behaviour and properties of quantum systems. The observer's choice of measurement apparatus or the act of measuring itself collapses the superposition of states into a well-defined outcome. This implies that the observer plays an active role in shaping the observed reality and that the observed system and the observer are entangled in a complex relationship. Also, whether consciousness plays a role in this issue is still unresolved. Currently, the most accepted explanation is the decoherence theory, as previously discussed. In the future, it will be intriguing to see if modern theories of consciousness, such as those involving integrated information, will play a significant role in addressing the measurement problem.

The principles of complementarity and wave-particle duality have not only expanded our understanding of the quantum world but also paved the way for practical applications and technologies. Quantum technologies, such as quantum computing and quantum cryptography, harness the unique properties of quantum systems, including superposition and entanglement, to perform calculations and secure communication. These technologies rely on the principles of complementarity and wave-particle duality to manipulate and control quantum states for specific purposes.

To summarise, the principles of complementarity and wave-particle duality are fundamental to quantum mechanics, profoundly reshaping our comprehension of quantum-level phenomena. Beyond enhancing our understanding of quantum mechanics, complementarity and wave-particle duality are pivotal to advancing quantum technologies, which hold promise for transformative developments in computing and communication.

A Non-Dual and Complementary World

Non-duality and complementarity are profound concepts emerging from different philosophical and scientific domains. Non-duality, often explored in spiritual and philosophical traditions, emphasises the inherent unity and interconnectedness of all things. Complementarity, arising in scientific disciplines such as quantum mechanics, highlights the coexistence of seemingly contradictory or mutually exclusive aspects. While distinct, these concepts share potential connections and insights that can enrich our understanding of reality.

1. Challenging Dualistic Thinking

Both non-duality and complementarity challenge dualistic thinking and invite us to transcend limited perspectives. Non-duality asserts that there is no inherent separation between the self and the world, while complementarity recognises the interplay of seemingly opposite aspects. Both emphasise the underlying unity of diverse manifestations, encouraging us to move beyond binary distinctions towards a more holistic and inclusive understanding of reality. This recognition allows us to embrace the richness and complexity of existence without imposing rigid boundaries.

2. Illusory Nature of Separation

Non-duality suggests that individuality and separateness are constructs of the ego, obscuring a deeper unity that permeates all things. Complementarity shows that observing or measuring quantum systems inherently involves a subjective interaction, indicating that the separation between observer and observed is not absolute. Both perspectives highlight the illusory nature of separation and encourage us to question perceived boundaries, recognising the interconnectedness that transcends apparent divisions. They challenge us to embrace the coexistence of seemingly contradictory aspects, expanding our capacity to hold paradoxes and transcend linear thinking.

3. Practical Insights

Non-duality and complementarity offer practical insights into our lives. Non-duality encourages cultivating a sense of oneness and compassion, transcending the illusion of separateness. Complementarity helps us appreciate the interplay of complementary aspects, fostering a more holistic understanding of complex issues. These perspectives invite us to approach situations with an inclusive and flexible mindset, recognising the interconnectedness of all beings and phenomena.

In summary, by reflecting on the connections between non-duality and complementarity, we can enrich our understanding of reality and our place in the interconnected web of existence. Both concepts challenge dualistic thinking and reveal interconnectedness within diversity. Embracing unity within diversity and acknowledging the illusory nature of separation can cultivate a more inclusive and compassionate worldview. This expanded perspective allows us to move beyond limited ways of thinking and experiencing the world, opening ourselves to greater possibilities and a deeper understanding of reality.

The Significance of Paradoxes in Trika Shaivism and Quantum Mechanics

Paradoxes—seemingly contradictory statements or situations that resist easy resolution—have long fascinated human thought and inquiry. They challenge our conventional understanding of reality and encourage the exploration of deeper truths. Both Trika Shaivism and quantum mechanics embrace the significance of paradoxes within their realms of inquiry, offering valuable insights into the interconnectedness of philosophical and scientific perspectives.

In Trika Shaivism, paradoxes, or antinomies, are recognised as vital for transcending ordinary dualistic thinking and approaching a deeper understanding of reality. Rather than seeking to resolve these paradoxes, Trika Shaivism encourages individuals to dwell in their tension, exploring the deeper truths that emerge from the interplay of seemingly opposing concepts. Notable paradoxes in Trika Shaivism include:

1. Identity and Difference

Trika Shaivism acknowledges that all existence is rooted in the absolute unity of consciousness (Shiva), yet within this unity, infinite diversity and individuality arise. Practitioners are invited to embrace the coexistence of oneness and multiplicity, moving beyond binary thinking.

2. Manifestation and Non-Manifestation

The tradition explores the paradox of reality existing in constant flux, with phenomena manifesting and dissolving, while simultaneously recognising the unchanging, transcendental essence that underlies all change. This paradox encourages contemplation of the interplay between the manifest and non-manifest realms.

Quantum mechanics also presents paradoxes that challenge classical intuitions and offer insights into the nature of reality at the quantum level. These paradoxes include:

1. Wave-Particle Duality

This phenomenon suggests that particles such as electrons and photons exhibit both wave-like and particle-like behaviours, depending on the experimental context. This challenges the classical notion of objects having distinct properties and calls for a more nuanced understanding of the quantum realm.

2. Uncertainty Principle

Formulated by Werner Heisenberg, this principle asserts that certain pairs of physical properties, such as position and momentum, cannot be simultaneously measured with arbitrary precision. This introduces fundamental uncertainty into the fabric of reality and challenges classical determinism.

3. Quantum Entanglement

This paradox describes a situation where particles become correlated such that the measurement of one particle instantaneously affects the state of another, regardless of distance. This non-local correlation may appear to violate relativistic constraints, in applications such as quantum teleportation, if it is just information and not 'useful' information that is taken into account. It points to a deeper interconnectedness among quantum systems.

The presence of paradoxes in both Trika Shaivism and quantum mechanics highlights the limitations of conventional binary thinking and invites a more holistic understanding of reality. Paradoxes challenge us to move beyond simple categorisations and recognise the interplay of seemingly opposing concepts. They serve as gateways to deeper truths, expanding our consciousness and encouraging us to explore the mysteries of existence with humility and curiosity. By embracing the paradoxes inherent in these traditions, we can deepen our appreciation for the intricate and interconnected nature of existence, transcending conventional boundaries and broadening our understanding of reality.

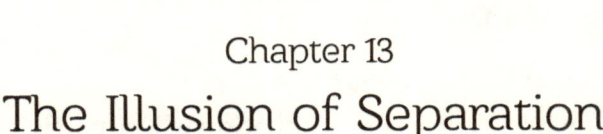

Chapter 13
The Illusion of Separation

The perception of separation—the idea that we exist as distinct beings separate from the world around us—is deeply ingrained in our daily experience and cultural narratives. However, both metaphysics and science challenge this perception, suggesting that separation is an illusion. This chapter explores the concept of the illusion of separation from metaphysical and scientific perspectives, examining how these domains illuminate the interconnected nature of reality and the implications of recognising this illusion for our understanding of ourselves and the world.

Transcending the Illusion of Separation: Insights from Philosophy and Science

Metaphysics, the branch of philosophy exploring the fundamental nature of reality, offers profound insights into the illusion of separation. Traditions such as Advaita Vedanta, Buddhist philosophy, and various indigenous wisdom teachings emphasise the interconnectedness and unity of all things. They suggest that the perception of separation arises from the limitations of our conditioned minds, which create a sense of individuality. According to these perspectives, the true nature of reality is non-dual, where all apparent divisions are illusory.

Science, particularly in fields like physics, biology, and neuroscience, also challenges the perception of separation. In physics, quantum entanglement reveals that particles can become intrinsically linked regardless of spatial separation, suggesting an underlying interconnectedness at the fundamental level of reality. Biology's study of ecosystems highlights the interdependence of different organisms and their environment. Neuroscience explores how the brain constructs the sense of self, revealing that our notion of a separate self is a product of neural processes.

Moreover, both metaphysical and scientific perspectives not only draw attention to the role of perception and consciousness in shaping our experience of separation but also offer practices to transcend the illusion of separation. These practices can help us shift our perspective, expand our consciousness, and experience a deeper sense of connectedness with ourselves, others, and the world. Recognising the illusion of separation has significant implications for our personal and global perspectives. It invites us to cultivate interconnectedness, compassion, and empathy towards all beings, encouraging us to transcend the limitations of our egos and embrace a broader understanding of ourselves as interdependent with the world. From a global perspective, it challenges divisive narratives and fosters a sense of global citizenship and environmental responsibility.

In conclusion, the illusion of separation, while deeply ingrained in our perception and cultural narratives, is challenged by both metaphysical and scientific perspectives. By exploring practices from metaphysics and science, we can transcend the illusion of separation and experience a deeper sense of unity with all existence.

Exploring Unity in Trika Shaivism: Insights into Interconnectedness

Trika Shaivism offers profound insights into the nature of existence and the illusion of separation. This philosophical tradition recognises that the perception of separation is a fundamental misunderstanding of reality and emphasises the unity of all existence. This section will explore Trika Shaivism's perspective on the illusion of separation, its teachings on the interconnectedness of all existence, and the practical implications of this recognition. The concept of spiritual oneness and non-separability is captured in verse 1.4.4 of the *Īśvarapratyabhijñākārikā*, which explains that experiences are not remembered as isolated objects. Instead, they are recalled as integral parts of one's Self, emphasising that all experiences are unified within the Self.

नैव ह्यनुभवो भाति स्मृतौ पूर्वोऽर्थवत्पृथक् ।
प्रागन्वभूवमहमित्यात्मारोहणभासनात् ॥

Trika Shaivism asserts that the perception of separation is an illusion arising from our conditioned minds and limited perspectives. It recognises that the sense of individuality stems from identification with the ego and belief in a separate self. This illusion of separation leads to suffering and limits our understanding of reality's true nature. By recognising the illusory nature of separation, individuals can transcend limited identifications and experience a deeper sense of interconnectedness with all beings and the world.

Central to Trika Shaivism is the concept of Shiva consciousness, understood as the ultimate reality from which all beings and phenomena arise. The apparent diversity and multiplicity of existence are seen as manifestations of Shiva consciousness expressing itself in various forms. Recognising this interconnectedness is seen as a means to expand our consciousness and cultivate unity and compassion towards all beings. By acknowledging our interconnectedness, individuals can move beyond a narrow sense of self and embrace a broader understanding of themselves as part of a greater whole.

To transcend limited identifications that perpetuate the illusion of separation, Trika Shaivism offers practices such as meditation, self-inquiry, and devotion. These practices enable individuals to cultivate awareness of their conditioned thought patterns and dissolve ego identification. By doing so, individuals can expand their consciousness and recognise their true nature as expressions of divine consciousness, experiencing a profound shift in their perception of themselves and the world.

Furthermore, the recognition of interconnectedness has practical implications for both individuals and the world. On a personal level, it invites the cultivation of compassion towards all beings and fosters a sense of unity, empathy, and interconnectedness, enhancing personal well-being and relationships. On a global scale, it calls for a shift in collective consciousness, encouraging a move beyond divisive narratives towards a more inclusive and holistic perspective. This recognition promotes work towards social justice, environmental sustainability, and collective well-being.

Beyond Locality: Exploring the Quantum Interconnectedness

Quantum entanglement is a fascinating phenomenon in quantum physics that refers to the correlation and interdependence of quantum systems, where the states of two or more particles become inseparably linked. When particles are entangled, their properties become intertwined, and the state of one particle cannot be described independently of the others. This means that the behaviour of one particle is instantaneously connected to the behaviour of another, regardless of the distance between them, challenging the classical notion of locality and separability.

Experimental evidence has consistently confirmed the reality of quantum entanglement. The Bell test experiments, which measure the correlation between properties of entangled particles, consistently violate Bell's inequalities, providing strong evidence for non-local correlations. Other experiments, such as delayed-choice experiments and quantum teleportation, further substantiate the existence of entanglement and its impact on quantum systems.

To better understand quantum entanglement, various theoretical frameworks provide explanations for this phenomenon. The Copenhagen interpretation suggests that entangled particles exist in a superposition of all possible states until measured, at which point the entangled state collapses. The many-worlds interpretation posits that entangled particles split into multiple parallel universes, each corresponding to a different measurement outcome. These frameworks provide different perspectives on the nature of entanglement.

The implications of quantum entanglement extend to various applications and technologies. It forms the basis for quantum communication protocols such as quantum teleportation, cryptography, and key distribution. Entanglement also plays a crucial role in quantum computing, where entangled qubits can perform complex calculations more efficiently than classical computers.

Beyond practical applications, quantum entanglement has broader implications for our understanding of reality. It challenges classical intuitions about separability and the existence of independently existing entities. The interconnectedness suggested by entanglement extends beyond particles and has been hypothesised to influence biological and macroscopic systems. This phenomenon pushes the boundaries of our understanding and raises profound questions about the fundamental nature of reality.

To conclude, quantum entanglement reveals the remarkable interconnectedness of quantum systems, challenging notions of separability and locality, and beckons us to explore the intricate web of connections that underlie the quantum fabric of the universe.

The Unified Perspectives of Trika Shaivism and Quantum Physics

As we have seen so far, Trika Shaivism and quantum mechanics offer profound insights into the nature of reality, challenging our perception of separation. Both perspectives recognise the limitations of our perception and emphasise the interconnectedness and unity underlying apparent divisions. Let's look at some of the important takeaways from these perspectives.

1. **Shared Implications**

Trika Shaivism teaches that all beings and phenomena are expressions of a greater whole, interconnected with divine consciousness. Quantum mechanics reveals entanglement and non-local correlations between quantum systems, suggesting deep interconnectedness at the fundamental level of reality. Both invite us to transcend dualistic thinking, embrace unity within diversity, and move beyond limited identifications.

2. **Role of Perception and Consciousness**

Trika Shaivism suggests that our perception of separation arises from conditioned minds and ego identification. It emphasises expanding consciousness and dissolving limited identifications to realise the underlying unity of existence. Quantum mechanics reveals that measurement or observation influences the behaviour of quantum systems, suggesting an active role of consciousness in shaping observed reality. This shared understanding highlights the connection between perception, consciousness, and the illusion of separation.

3. **Unity Within Diversity**

Trika Shaivism emphasises a fundamental unity at the core of all beings and phenomena, describing it as divine consciousness from which all manifestations arise. Quantum mechanics, through entanglement and non-separability, reveals interconnected quantum systems with non-local correlations, suggesting that apparent diversity is a manifestation of a unified whole. Both perspectives invite us to embrace the complexity of existence while recognising fundamental unity.

4. Practical Implications and Applications

Both perspectives encourage cultivating interconnectedness, compassion, and empathy towards all beings. They promote expanding consciousness, dissolving limited identifications, and recognising underlying unity. This recognition has implications for personal well-being, relationships, and interactions with the world, encouraging us to move beyond divisive narratives and foster unity and harmony. The shared understanding can also apply to consciousness studies, quantum technologies, and holistic approaches to science and spirituality.

5. Transcending Limitations

Trika Shaivism emphasises transcending limited identifications through spiritual practices. Quantum mechanics challenges classical notions of separability and locality, revealing entanglement and non-local correlations. This shared understanding invites us to transcend dualistic thinking, embrace unity within diversity, and recognise the unity pervading all aspects of reality, with profound implications for our perception of self, relationships, and our place in existence.

In conclusion, to better understand the concepts of separability and unity in both Trika Shaivism and quantum physics, it is essential to understand their different approaches. Trika Shaivism focuses on achieving spiritual unity through inner transformation, while quantum physics seeks to provide empirical evidence of unity in the physical world. Despite these differing methods, both perspectives converge on the idea that separateness is an illusion and that true understanding comes from recognising the underlying interconnectedness. Integrating these views reveals how Trika Shaivism's spiritual insights and quantum physics' empirical findings complement each other. Together, they offer a more comprehensive understanding of reality, blending spiritual and scientific perspectives to uncover the unity that underlies all existence.

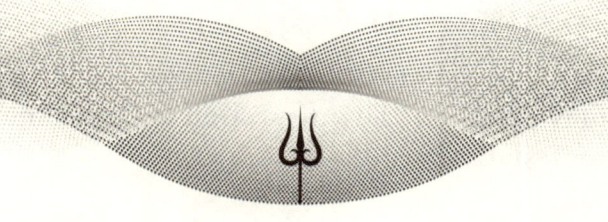

Chapter 14
The Observer's Role in Manifesting Reality

The insight that observations influence reality marked a fundamental shift in our understanding of reality and its evolution in early 20th-century science, suggesting that the universe is more participatory than we might expect. The observer effect is an integral aspect of quantum physics and it highlights the profound influence of observation on the behaviour of quantum systems. It suggests that the act of observation or measurement affects the properties and behaviour of particles. This raises intriguing questions about the role of consciousness and the observer in shaping reality.

Parallel to the scientific observer effect, the concept of manifestation, explored in metaphysical and spiritual traditions, emphasises the power of thought and intention in shaping one's reality. These traditions, with their long history, suggest a deep-rooted connection between our perceptions and the nature of reality. This chapter will explore the relationship between the observer effect and manifestation, delving into the scientific and metaphysical aspects of these concepts and their implications for our understanding of reality and conscious creation.

Conscious Creation: The Intersection of Quantum Physics and Manifestation

In quantum physics, the observer effect refers to the phenomenon where the act of observation influences the behaviour and properties of quantum systems. This suggests that the act of observation has a direct impact on the manifestation of a particular reality. The observer effect raises intriguing questions about the role of consciousness and the observer in shaping reality. Some interpretations of quantum mechanics suggest that the observer's consciousness interacts with `the quantum system, leading to the manifestation of a specific outcome. This viewpoint challenges the traditional notion of an objective reality independent of observation and highlights the intimate relationship between consciousness and the observed world.

On the other hand, the concept of manifestation emphasises the power of thought and intention in creating one's reality. It suggests that individuals have the ability to shape their experiences and materialise their desires through focused intention and belief. This process is believed to influence the energetic and vibrational aspects of reality, attracting the desired experiences into one's life. Manifestation practices often involve visualisation, affirmations, and aligning one's thoughts and emotions with the desired outcome.

Therefore, we can see that though the observer effect in quantum physics focuses on the behaviour of quantum systems, it shares intriguing parallels with the concept of manifestation. Both concepts recognise the profound influence of consciousness and intention on the nature of reality. The relationship between the observer effect and manifestation challenges our traditional understanding of reality as something objective and independent of observation and suggests that consciousness and intention play integral roles in shaping our experiences and the manifestation of the world around us. This viewpoint raises profound questions about the nature of reality, the interplay between consciousness and the external world, and the potential for conscious creation.

Beyond theoretical knowledge, these perspectives also offer practical applications in diverse areas of everyday life. By recognising the influence of consciousness and intention on reality, individuals can consciously create their experiences and work towards manifesting their desires. However, ethical considerations are crucial in manifestation practices. It is essential to exercise responsibility, integrity, and compassion in the process of conscious creation to ensure that desired outcomes align with the highest good for oneself and others.

To summarise, the observer effect and manifestation offer intriguing insights into the relationship between consciousness, observation, and reality. Exploring the parallels between these concepts invites us to reconsider our understanding of reality and the role of consciousness in shaping our experiences. Integrating scientific and metaphysical perspectives deepens our exploration of consciousness and its potential impact on manifesting our lives.

Conscious Creation: Trika Shaivism's Path to Manifesting Reality

Let's look at some of the key perspectives of Trika Shaivism on conscious manifestation:

1. **Consciousness as the Essence of Reality**

Trika Shaivism asserts that consciousness is the essence of reality and that our perception of reality is intricately connected to the quality and depth of our consciousness. It posits that the divine consciousness is the fundamental source from which all manifestations arise and that reality is not separate from consciousness but is a dynamic expression of it. Recognising consciousness as the essence of reality allows individuals to shift their focus from external circumstances to their inner realm. This realisation empowers them to influence their experiences through conscious awareness.

2. **Power of Thought and Intention**

Trika Shaivism emphasises the power of thought and intention in shaping reality. It teaches that our thoughts and intentions have a direct impact on the manifestation of our experiences. It suggests that by aligning our thoughts and intentions with the divine consciousness, we can attune ourselves to a higher vibration and attract positive and harmonious experiences into our lives. This understanding empowers individuals to take responsibility for their thoughts and cultivate positive and uplifting mental states.

3. **Awareness and Mindfulness**

Awareness and mindfulness also play a crucial role in Trika Shaivism's understanding of conscious manifestation. It teaches that by cultivating awareness and being fully present in each moment, individuals can tap into the creative power of consciousness. By becoming aware of their thoughts, emotions, and beliefs, individuals can consciously choose those that align with their desired experiences. Mindfulness practices such as meditation, self-inquiry, and contemplation are integral to Trika Shaivism's teachings, as they help individuals deepen their awareness and access the transformative potential of consciousness.

4. **Transcending Limiting Beliefs**

Trika Shaivism states that limiting beliefs and egoic patterns can hinder the manifestation of desired realities. These patterns arise from conditioning and societal influences, and they create a sense of separation and limitation. Trika Shaivism encourages individuals to transcend these limiting beliefs and egoic patterns by cultivating self-awareness and realising their inherent divinity. By recognising the illusory nature of the ego and identifying with the higher self, individuals can transcend limitations and manifest their true potential.

5. **Aligning with Divine Will**

Trika Shaivism emphasises the importance of aligning our intentions with the divine will. It teaches that true manifestation occurs when our intentions are in harmony with the greater cosmic plan. This alignment requires surrendering the egoic desire for control and trusting in the wisdom and intelligence of the divine consciousness. By aligning with the divine will, individuals can co-create with the creative power of consciousness and manifest experiences that are in accordance with their highest good.

6. Ethical Conduct and Service

Trika Shaivism emphasises the importance of ethical conduct and service to others in conscious manifestation. It teaches that manifestation is not solely for personal gain but should also consider the welfare of others and the greater good. It further encourages individuals to cultivate compassion, kindness, and selflessness in their thoughts, words, and actions. By serving others and contributing to the well-being of the world, individuals create a positive karmic cycle that supports their own conscious manifestation.

In conclusion, Trika Shaivism's emphasis on consciousness in manifesting reality offers profound insights and practical tools for the conscious creation of experiences. By recognising consciousness as reality's essence and aligning thoughts, intentions, and actions with divine consciousness, individuals can tap into consciousness's creative power. These teachings remind us of our inherent transformative potential and the responsibility to cultivate awareness, transcend limitations, and manifest realities reflecting our inherent divinity.

The Observer Effect: Shaping Quantum Reality through Observation

In quantum mechanics, the observer effect refers to the phenomenon where observation or measurement affects quantum systems' behaviour and properties. When an observation is made, the wave function collapses to a specific state, corresponding to the observed outcome. This collapse of the wave function leads to the manifestation of a definite value for the observed property and eliminates the superposition of multiple states. Thus, the observer effect highlights the inherent connection between the act of observation and the behaviour of quantum systems.

The observer effect raises profound questions about the nature of reality and the role of observation in shaping our understanding of the quantum world. It suggests that our observations and measurements actively participate in the creation of reality. The collapse of the wave function implies that observation brings about the manifestation of a particular reality, highlighting an intimate relationship between consciousness and the quantum world. This viewpoint challenges traditional notions of objective reality independent of observation. Moreover, the implications of the observer effect extend beyond physics, prompting us to reconsider the role of consciousness in shaping our experiences and the fundamental nature of the universe we inhabit.

Bridging Spirituality and Science: Observer Effect in Trika Shaivism and Quantum Mechanics

The observer effect in quantum mechanics and the role of consciousness in Trika Shaivism present intriguing parallels, despite originating from different fields. This section delves into the underlying principles, implications, and shared insights that emerge from these seemingly disparate domains.

1. Parallels in Observer Role

    Both perspectives recognise the observer's significance in perception and reality creation. Trika Shaivism suggests that consciousness and awareness play a vital role in shaping subjective experiences. By observing thoughts, emotions, and beliefs, individuals can transcend conditioned patterns, leading to a profound understanding of their true nature and universal interconnectedness.

2. Non-Duality Concept

Trika Shaivism teaches that reality's true nature is non-dual, with apparent divisions being illusory. Similarly, quantum mechanics describes the superposition of quantum states. This parallel suggests that the observer effect and wave function collapse in quantum mechanics can be understood as a transition from a superposition of possibilities to a specific manifestation, akin to the transition from non-duality to duality in Trika Shaivism.

3. Convergence on Consciousness

Both quantum mechanics and Trika Shaivism highlight the participatory nature of the universe. In quantum mechanics, measurement is essential to the collapse of the wave function, making the act of observation fundamental to the reality that emerges. Similarly, Trika Shaivism emphasises the observer's role in shaping the perception of reality. Both frameworks acknowledge the profound connection between the observer and the manifestation of the observed world.

The parallels between the observer effect in Trika Shaivism and quantum mechanics highlight shared insights and interconnectedness between seemingly disparate fields. They challenge the notion of an objective, independent reality and emphasise the interplay between observer and observed. Both perspectives invite us to consider our active role in shaping experiences through consciousness and awareness.

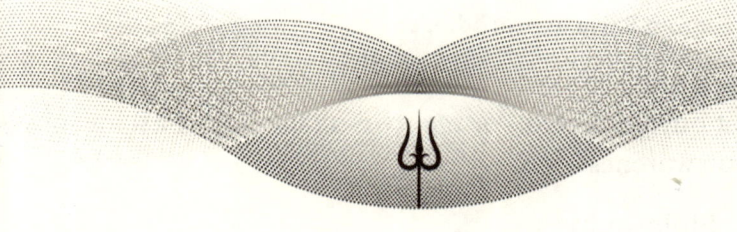

Chapter 15
The Nature of Reality

Reality is multifaceted and may extend beyond our conventional understanding, existing outside familiar dimensions. The nature of reality has been a subject of profound inquiry across metaphysics, philosophy, and science. Each discipline offers unique perspectives for exploring and understanding the fundamental nature of existence. Let's look at the different views on reality, examining key concepts, exploring their interconnections, and discussing the implications for our understanding of ourselves and the world around us.

1. Metaphysics

Metaphysics explores the fundamental nature of reality beyond the physical realm. It suggests that reality encompasses unseen dimensions and realms, extending beyond the observable and tangible. Some metaphysical frameworks propose a transcendent reality, such as a divine or spiritual realm, as the ultimate source of existence. Concepts like consciousness, energy, and interconnectedness are central to metaphysical perspectives, highlighting the interplay between material and immaterial realms. These perspectives often also posit that our perception of reality is shaped by consciousness and awareness.

2. Philosophy

Philosophy offers various perspectives on the nature of reality, with different philosophical traditions providing unique insights. Idealism, for instance, posits that reality is fundamentally mental or conceptual in nature, with the physical world being a product of our perceptions and mental constructs. In contrast, materialism suggests that reality is entirely composed of physical matter and can be understood through scientific investigation. Other philosophical perspectives, such as dualism, propose a fundamental duality between mind and matter or between the physical and spiritual aspects of reality. Philosophical perspectives also delve into questions regarding existence, identity, causality, and the nature of knowledge, offering diverse interpretations and frameworks for understanding reality.

3. Science

Science employs rigorous empirical methods to understand reality, emphasising observation, experimentation, and theory formulation to explain natural phenomena. Scientific inquiry has advanced our understanding from subatomic particles to galaxies. It proposes that reality is governed by laws across physics, chemistry, biology, and other disciplines, prioritising objectivity, empirical evidence, and reproducibility to establish knowledge.

In conclusion, all these diverse perspectives offer profound insights into ourselves and the world. Metaphysical perspectives from traditions like Trika Shaivism emphasise the profound interconnectedness of all things, proposing that we are integral parts of a greater whole in an interconnected web of existence. Philosophical perspectives encourage reflection on consciousness, identity, and the limits of human understanding. Scientific frameworks provide a systematic, evidence-based approach to studying the physical world and advancing technology. And while metaphysics, philosophy, and science are often viewed as distinct fields, they share intersections and interdependencies. Metaphysics provides overarching frameworks that encompass philosophical and scientific inquiries, offering insights into consciousness, interconnectedness, and the fundamental nature of reality. Philosophy critically analyses and explores concepts, contributing to metaphysical and scientific theories. Science, on the other hand, offers empirical data, observations, and experimental evidence that inform both metaphysical speculation and philosophical inquiry. This interplay between disciplines facilitates a more comprehensive exploration of the nature of reality. Integrating these perspectives and connecting their insights has the potential to deepen our understanding of the complexities and nuances inherent in the nature of reality. Achieving this integration requires an open-minded approach, a readiness to explore diverse viewpoints, and an acknowledgement of the unique contributions each discipline offers to our pursuit of knowledge.

Ultimately, the nature of reality stands as a vast and multifaceted subject that has long intrigued philosophers, metaphysicians, and scientists alike. Embracing insights from these disciplines allows us to develop a more holistic and enriched understanding of reality, empowering us to explore our place within it and contemplate the mysteries that extend beyond our current comprehension.

From Quantum Particles to Vibrating Strings: Exploring Fundamental Physics

In physics, various theories such as quantum mechanics, quantum field theory, the Standard Model, and string theory offer distinct perspectives on the fundamental nature of the universe. In this section, we will explore key concepts from these theories that shed light on the nature of reality.

1. **Quantum Mechanics**

   Quantum mechanics has revolutionised our comprehension of the microscopic world through the introduction of wave-particle duality. This concept challenges classical perspectives by suggesting that particles can display both wave-like and particle-like behaviours. It highlights the probabilistic nature of quantum systems and is encapsulated by the uncertainty principle, which restricts our ability to precisely determine a particle's position and momentum simultaneously. These insights underscore inherent uncertainty at the quantum level and highlight the intrinsic limits of our knowledge.

2. **Quantum Field Theory and the Standard Model**

   Quantum field theory extends quantum mechanics by introducing fields as the foundational entities of nature. According to this framework, particles are interpreted as excitations of underlying fields that extend throughout space and time. These fields, like the electromagnetic field or the Higgs field, are quantised and exist only in discrete packets or quanta. QFT also offers a mathematical framework to elucidate how these fields behave and interact, serving as the foundation for understanding fundamental forces and particles.

The Standard Model is a successful theory within the framework of quantum field theory. It describes fundamental particles constituting matter and the three fundamental forces (electromagnetic, weak, and strong interactions) governing their behaviours. The discovery of the Higgs boson, linked to the Higgs field, validated how particles acquire mass. This model offers profound insights into the particle-level structure of reality, encompassing all known particles and their interactions.

3. String Theory

String theory is a theoretical framework that aims to unify quantum mechanics and general relativity. It suggests that the fundamental constituents of the universe are not particles but tiny, vibrating strings. These strings can vibrate in different modes, generating the diverse particles observed in nature. An essential aspect of string theory is its requirement for extra dimensions beyond the familiar three spatial dimensions. These extra dimensions are hypothesised to be compactified or curled up at scales currently beyond our observational reach. String theory presents a compelling vision of reality that includes additional dimensions and aims to unify the fundamental forces of nature. However, it remains a subject of active research with many unresolved issues.

Quantum mechanics, quantum field theory, the Standard Model, and string theory provide valuable insights into the nature of reality, yet they also pose significant challenges. Quantum mechanics, with its probabilistic nature, challenges classical notions of determinism and predictability. Quantum field theory introduces fields as fundamental entities, expanding our understanding of nature's building blocks. The Standard Model successfully describes known particles and their interactions but leaves mysteries like dark matter and unifying gravity unresolved. Together, these theories offer complementary perspectives on reality, challenging classical intuitions and revealing insights into quantum systems' probabilistic behaviour, the foundational role of fields, particle interactions, and the potential for extra dimensions. Researchers continue to explore and seek a unified framework that can explain all phenomena, aiming for deeper insights into the fundamental nature of the cosmos.

Trika Shaivism's Perspective on Reality

Trika Shaivism, with its modes of spiritual inquiry, has a distinct perspective on the fundamental nature of reality. In this section, we will delve into the core tenets of Trika Shaivism, exploring its profound views on the nature of reality and consciousness. We will examine how consciousness is considered the fundamental essence of all existence, the cosmic dance of Shiva, the dynamic role of Shakti, the concept of maya, and the pivotal idea of spanda. Through these principles, Trika Shaivism offers a rich framework for understanding the interplay between the manifest world and the underlying divine consciousness.

1. Consciousness as the Foundation

Trika Shaivism asserts that consciousness is the fundamental essence of all existence. It posits that reality is not separate from consciousness but a dynamic expression of it. The ultimate reality, known as Shiva, is understood as pure consciousness that transcends conceptual understanding. From this absolute consciousness arises the entire spectrum of manifestation, encompassing both material and immaterial aspects of existence.

2. The Cosmic Dance of Shiva

Trika Shaivism describes the process of manifestation as the cosmic dance of Shiva. This dance represents the continuous interplay between consciousness and manifestation, as Shiva projects and withdraws the universe. The manifestation phase involves the unfolding of the universe with its countless forms, energies, and experiences, while the withdrawal phase entails the dissolution of these manifestations back into the source consciousness, reflecting the perpetual cycle of creation and dissolution.

3. Role of Shakti

Trika Shaivism recognises the active role of Shakti, the creative power of consciousness, in the process of manifestation. Shakti represents the dynamic aspect of consciousness that gives rise to the diversity of forms and experiences. It is the divine feminine energy that permeates all of existence. The interplay between Shiva and Shakti is seen as the driving force behind the continuous unfolding and transformation of reality.

4. Maya and Illusion

Trika Shaivism acknowledges the presence of maya, the illusion of separation and duality, in the realm of manifestation. Maya creates the appearance of individuality, veiling the underlying unity of consciousness. It gives rise to the perception of separate entities, experiences, and limitations. Recognising and transcending maya is considered essential for realising the non-dual nature of reality and experiencing the unity of all existence.

5. Concept of Spanda

A key concept in Trika Shaivism is spanda, which refers to the pulsation or vibration of consciousness. Spanda represents the inherent dynamism and vibrancy of consciousness, manifesting as the creative impulse and the movement of energy. It is the underlying rhythm that gives rise to the continuous flux and transformation of reality. Attunement to spanda allows individuals to align themselves with the inherent creative power of consciousness and participate consciously in the co-creation of their experiences.

All in all, Trika Shaivism invites individuals to transcend the illusion of separation, deepen their awareness of the interconnection of all existence, and align their consciousness with the inherent creative power of reality. Through this understanding, individuals can embark on a transformative journey of self-discovery and co-creation, experiencing the profound beauty and unity that underlies the fabric of existence.

Can Reality be Quantised?

Quantum Mechanics raises as many questions as it answers, for instance, vis-à-vis the conceptual divergences in understanding the quantum wave-function. Psi-ontic scholars regard it as an actual entity, while psi-epistemic scholars regard it as a descriptor. Rife with space for interpretations, we must look closely at the different aspects of the quantum realm to understand it better.

1. Probabilistic Nature of Quantum Mechanics

Superposition and wave-particle duality underscore the probabilistic framework underlying the quantum world. This implies that particles and systems can exist in multiple states simultaneously and can only be described in terms of probabilities. Additionally, the Copenhagen interpretation, proposed by Niels Bohr and Werner Heisenberg, also embraces the probabilistic nature of quantum mechanics and introduces the concept of wave function collapse upon measurement. This interpretation emphasises the importance of the observer and separates the classical world of macroscopic objects from the quantum realm of microscopic particles. Alternatively, the many-worlds interpretation, put forth by Hugh Everett III, offers a different explanation for the probabilistic nature of quantum mechanics. According to this interpretation, every possible outcome of a quantum measurement actually occurs in different branches or universes, resulting in a multitude of parallel realities.

In addition to the Copenhagen interpretation and the many-worlds interpretation, there are various other interpretations of quantum mechanics, each offering a different perspective on the probabilistic nature of reality. These include the transactional interpretation, the consistent histories interpretation, and the objective collapse theories. These interpretations have significant implications not only for our understanding of quantum mechanics but also for broader philosophical questions about the nature of reality, free will, and the role of the observer.

2. Non-Locality and Unity

In the realm of quantum mechanics, two intriguing approaches offer alternative perspectives on the nature of reality: Bohmian mechanics and entanglement. While both approaches diverge from the conventional interpretations of quantum mechanics, they provide unique insights into the fundamental fabric of the universe. Bohmian mechanics, also known as the de Broglie-Bohm theory or the pilot-wave theory, is a deterministic interpretation of quantum mechanics. It was formulated by physicist David Bohm as an alternative to the probabilistic nature of standard quantum mechanics. In Bohmian mechanics, particles possess definite positions and trajectories, guided by a hidden variable known as the quantum potential or the pilot wave. The wave function still plays a central role in describing the probabilistic behaviour of quantum systems, but it no longer represents a complete description of reality. Instead, particles are influenced by both their own trajectories and the influence of the quantum potential, giving rise to the observed statistical distribution of measurement outcomes. This interpretation maintains determinism by introducing hidden variables, which are not directly observable.

Another striking feature of Bohmian mechanics is its inherent non-locality. In standard quantum mechanics, the collapse of the wave function is considered an instantaneous process, seemingly violating the limits of causality. However, in Bohmian mechanics, the apparent non-locality arises from the influence of the quantum potential, which can instantaneously affect the behaviour of particles across vast distances. A similar resonance can be found in the perspective of entanglement which suggests that reality extends beyond localised entities to form an interconnected web of entangled particles. This view challenges our classical intuitions and notions of separability, highlighting that reality is not composed of independent entities but rather an intricate tapestry of entangled relationships. In this framework, entangled particles are not separate entities but aspects of a unified whole. This perspective aligns with non-dualistic philosophies, implying a profound interconnectedness at the fundamental level of reality.

Bohmian mechanics and the entanglement perspective offer alternative viewpoints on the nature of reality within quantum mechanics. While differing in their formalisms and mathematical descriptions, both approaches share common themes that challenge our conventional understanding of reality. Both emphasise non-locality and interconnectedness, albeit from different angles—Bohmian mechanics provides a deterministic framework incorporating non-local influences, whereas the entanglement perspective highlights the inseparable nature of entangled particles. These complementary perspectives offer new ways of grappling with the mysteries of quantum mechanics and provide alternative interpretations of reality, provoking deeper philosophical and scientific inquiry into the fundamental nature of existence.

All in all, as we continue to explore the frontiers of quantum mechanics, these alternative perspectives provide valuable insights and open new avenues for understanding the enigmatic nature of reality.

Connections between Trika Shaivism and Quantum Mechanics' Perspectives on Reality

This section examines the fascinating parallels between Trika Shaivism and quantum mechanics in their views on the nature of reality. By examining these connections, we can gain insights into the fundamental nature of existence in both spiritual and scientific contexts.

1. Non-duality and Interconnectedness

Trika Shaivism emphasises non-duality, seeing ultimate reality as a unified whole and regarding apparent diversity as an illusion. Quantum mechanics also reveals non-dualistic nature through entanglement and superposition. These quantum phenomena reflect the interconnectedness and unity that Trika Shaivism espouses.

2. Observer Effect and Consciousness

Quantum mechanics' observer effect highlights the possible role of consciousness, as per certain interpretations, in shaping reality, suggesting observation affects quantum systems' behaviour and properties. This parallels Trika Shaivism's emphasis on consciousness as the foundation of existence. From Trika Shaivism's perspective, reality is a manifestation of consciousness, and the observer's consciousness influences perceived reality. Thus, we can see that both, albeit in specific strands of interpretation, recognise an inherent connection between consciousness and the nature of reality.

3. Wave-particle Duality and Manifestation

Wave-particle duality in quantum mechanics aligns with Trika Shaivism's understanding of manifestation. Particles exhibiting both wave-like and particle-like properties mirror Trika Shaivism's view of underlying reality (waves) giving rise to apparent diversity (particles). Both assert the dynamic interplay between unmanifest and manifest aspects of reality.

4. Uncertainty and Illusion

Heisenberg uncertainty principle in quantum mechanics posits limits on simultaneously knowing a particle's position and momentum, challenging determinate reality. Similarly, Trika Shaivism teaches that separation perception is an illusion created by the conditioned mind. Both emphasise understanding limitations and point to reality's interconnected and uncertain nature.

5. Timelessness and Eternal Consciousness

Trika Shaivism posits timeless, eternal consciousness transcending time and space limitations. This aligns with the quantum mechanical understanding of particles existing in superposition across time until measurement. Both suggest consciousness and reality exist beyond linear time constraints, with time perception as a mental construct.

Exploring the potential connections between Trika Shaivism's understanding of reality and quantum mechanics reveals intriguing parallels and shared principles. Concepts such as non-duality, the observer effect, wave-particle duality, uncertainty, and timelessness resonate deeply with both perspectives, suggesting an underlying unity between spiritual and scientific views of reality. While Trika Shaivism offers a metaphysical and experiential framework, quantum mechanics provides a mathematical and empirical description of the microscopic world. Integrating these perspectives invites us to contemplate the profound mysteries of existence and embrace a holistic understanding that spans from the inner dimensions of consciousness to the outer realms of quantum phenomena.

Guidelines for Thoughtful Integration of Trika Shaivism and Quantum Mechanics

Exploring connections and parallels between diverse fields of knowledge can be a fruitful endeavour if approached responsibly. Trika Shaivism, a profound philosophical and spiritual tradition, and quantum mechanics, a branch of physics, provide rich frameworks for understanding the nature of reality. However, it is essential to approach the task of drawing parallels between these two domains cautiously. In this section, we will explore why it is important to exercise caution when drawing parallels between Trika Shaivism and quantum mechanics. We will discuss potential pitfalls and limitations, and provide guidelines for a responsible and nuanced comparison of these complex and distinct perspectives.

1. Contextual Differences

Trika Shaivism and quantum mechanics emerge from vastly different cultural, historical, and epistemological contexts. Trika Shaivism is rooted in ancient Indian philosophy and spirituality, with a focus on metaphysics, consciousness, and self-realisation. Quantum mechanics, on the other hand, is a scientific discipline that emerged in the early 20th century, grounded in empirical observations, mathematical formalisms, and experimental verification. The differences in context imply that the language, concepts, and underlying assumptions of these domains may not align perfectly. Thus, caution is needed when drawing parallels to ensure that we do not impose an artificial compatibility.

2. Interpretive Variations

Both Trika Shaivism and quantum mechanics encompass diverse interpretations and perspectives. Trika Shaivism features multiple schools of thought and interpretations within its tradition. Similarly, quantum mechanics includes various interpretations like the Copenhagen interpretation, many-worlds interpretation, and pilot-wave theory, among others. These interpretive variations mean that scholars and practitioners within each field may emphasise different aspects and understandings. When exploring parallels, it is crucial to recognise and respect these variations, avoiding the imposition of singular interpretations on either domain.

3. Conceptual and Terminological Challenges

Trika Shaivism and quantum mechanics operate within distinct conceptual frameworks and terminologies. Trika Shaivism employs Sanskrit terminology and metaphysical concepts that may not directly correspond to the language and concepts of quantum mechanics. Attempting to draw direct parallels between these frameworks can lead to forced or inaccurate comparisons. Moreover, quantum mechanics utilises its own mathematical formalisms and technical language, which can be complex for non-specialists to grasp fully. A deep understanding of both domains is essential to avoid superficial or misleading connections.

4. **Epistemological Differences**

Trika Shaivism and quantum mechanics employ distinct methodologies for acquiring knowledge and understanding reality. Trika Shaivism emphasises subjective experience, meditation, and direct realisation, drawing insights from enlightened masters and transformative internal practices. In contrast, quantum mechanics follows a scientific approach relying on empirical observations, mathematical models, and experimental validation. These differing epistemological approaches underscore the importance of caution when seeking to integrate these domains. While subjective experiences provide valuable insights, they may not directly correlate with scientific observations and measurements.

Therefore, we can see that drawing parallels between Trika Shaivism and quantum mechanics demands careful consideration to prevent reductionism and misrepresentation. Issues such as reductionism may occur when complex ideas or systems are oversimplified or narrowed to fit into a different framework. Hence, it is crucial to respect the depth and complexity of both domains and avoid reducing them to simplistic comparisons. Misrepresentation can also arise from inaccurately applying concepts or principles, leading to misunderstandings or distortions. Ensuring a thorough and precise understanding of both Trika

Shaivism and quantum mechanics is vital to prevent misrepresentations.

Let's look at some of the key steps for integrating insights across these fields:

- Build a comprehensive understanding of both Trika Shaivism and quantum mechanics through thorough study and engagement with primary sources and experts.
- Appreciate and respect the cultural, historical, and epistemological distinctions between these two domains.
- Embrace the richness and complexity inherent in each tradition, avoiding oversimplification or reductionism.
- Rather than seeking direct parallels, focus on identifying common themes, principles, or philosophical inquiries shared by both.
- Explore the philosophical implications raised by both Trika Shaivism and quantum mechanics.
- Approach the integration of ideas with openness, curiosity, and critical inquiry.
- Foster respectful dialogue and diverse interpretations, encouraging a deeper exploration of these interconnected perspectives.

In summary, by fostering deep understanding, respecting the integrity of both domains and seeking common ground, we can engage in a meaningful dialogue that enriches our understanding of the nature of reality from diverse perspectives.

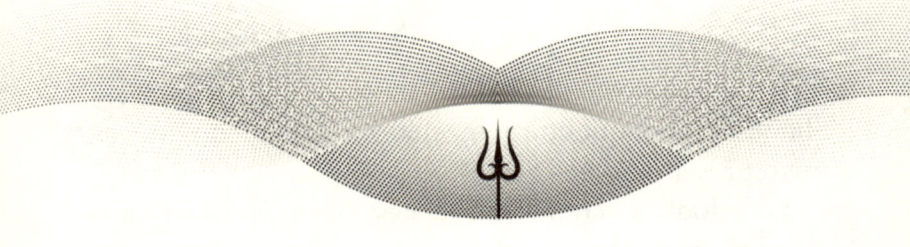

Chapter 16
Resonant Portraiture of Episteme

Epistemology delves into the study of knowledge—how it is acquired and its boundaries. It tackles fundamental questions such as: What constitutes knowledge? How do we attain it? What are its sources? And how do we assess the trustworthiness and accuracy of knowledge claims? This field encourages us to critically analyse the nature of knowledge and the methods by which we comprehend our surroundings. Furthermore, epistemology not only explores the nature of knowledge across various domains, including scientific, moral, and religious knowledge, but also considers the social and cultural factors that influence the acquisition and validation of knowledge.

This chapter will explore the key concepts and debates within epistemology as well as examine foundational questions about the scope and limits of human understanding, aiming to illuminate how we acquire knowledge and how different epistemological frameworks shape our perceptions of reality.

Exploring Epistemology: Examining the Foundations of Knowledge

One of the central questions in epistemology is the definition of knowledge itself. The traditional definition of knowledge, known as the justified true belief model, asserts that knowledge requires three conditions: belief, truth, and justification. According to this model, for something to be considered knowledge, it must be believed, it must be true, and there must be good reasons or evidence to support the belief. However, this definition has been subject to various criticisms and challenges, leading to the development of alternative theories of knowledge.

One such alternative is the reliabilist approach, which emphasises the reliability of the cognitive processes that lead to true beliefs. According to reliabilism, knowledge is not necessarily dependent on conscious justification but rather on the reliability of the cognitive mechanisms involved. This approach shifts the focus from the justification of beliefs to the reliability of the processes through which knowledge is acquired.

Another influential theory is the coherentist approach, which emphasises the coherence or consistency of beliefs within a broader system of beliefs. According to coherentism, knowledge is derived from the interconnectedness and coherence of beliefs rather than from individual justified beliefs. This approach challenges the foundationalist view that knowledge is built upon a set of indubitable or self-evident beliefs.

Epistemology also explores the sources and limits of knowledge. Empiricism, for instance, asserts that knowledge is primarily derived from sensory experience. Empiricists argue that our knowledge of the world is based on observations and inductive reasoning, where we generalise from specific instances to form general principles or laws. Rationalism, on the other hand, emphasises reason and rational thought as the primary sources of knowledge. Rationalists argue that certain truths can be known through reason alone, without relying on sensory experience.

Epistemology also examines the role of scepticism and the challenges it poses to our claims of knowledge. Scepticism raises doubts about the possibility of attaining certain knowledge and questions the reliability of our cognitive processes. It challenges our ability to distinguish between true beliefs and mere opinions, encouraging critical inquiry and prompting us to question the foundations of our knowledge claims. This, in turn, leads to a deeper understanding of the limitations of human knowledge.

In summary, epistemology is a dynamic field that continues to evolve and address new challenges. This epistemological inquiry stimulates critical thinking, fosters intellectual humility, and enriches our understanding of the complexities of human knowledge.

In the following section, we will explore the epistemological insights of Trika Shaivism and quantum mechanics, examining their shared perspectives on the nature of knowledge.

Direct Experiential Knowledge: Epistemological Insights from Trika Shaivism

Experiential knowledge is primary to Trika non-absolutism. Let's explore the guiding principles that underpin direct experiential knowledge in the Trika tradition. By examining these epistemological insights, we can gain a nuanced understanding of reality and its implications.

1. Non-absolutism in Trika Shaivism

One of the key principles within Shaivism, and Sanatana Dharma in general, is the concept of non-absolutism, which embraces pluralism, openness, and the recognition of reality's multifaceted nature. Non-absolutism, also known as anekantavada or multiplicity of viewpoints, asserts that reality is complex and cannot be fully captured by any single perspective or belief system. It acknowledges the inherent limitations of human understanding and encourages a humble and open-minded approach to spiritual and philosophical inquiry.

Non-absolutism in Sanatana Dharma fosters a spirit of pluralism, recognising and embracing the diversity of beliefs, practices, and paths to truth. Rather than advocating for a single absolute truth, non-absolutism encourages individuals to explore and honour the various paths and perspectives within the tradition, encompassing a wide array of philosophical systems, sects, deities, rituals, and spiritual practices.

This principle promotes an open-minded and inclusive approach to different viewpoints and perspectives. It acknowledges that truth is multifaceted and may be understood differently by various individuals or communities. This openness creates space for mutual understanding and the exchange of ideas which, in turn, allows for the coexistence of diverse beliefs and practices within the broader framework of Sanatana Dharma.

Non-absolutism also acknowledges the concept of relative truth. Meaning, it recognises that truth can be context-dependent and subjective and vary based on factors such as time, culture, and individual experience. This fosters a spirit of tolerance and respect for differing opinions, as it accepts that each perspective may contain a portion of the truth within its specific context.

In terms of personal spiritual growth, non-absolutism encourages individuals to engage in self-inquiry, critical thinking, and the exploration of different spiritual paths and practices. It invites seekers to move beyond rigid dogmas and fixed beliefs, allowing for personal evolution and growth. This approach fosters a dynamic and ever-evolving spiritual journey, where individuals can integrate diverse insights and experiences into their understanding of truth.

Non-absolutism also has broader implications for fostering harmony and understanding within society. By embracing pluralism and open-mindedness, it promotes dialogue, empathy, and mutual respect among individuals from different backgrounds and belief systems. It encourages the recognition that multiple viewpoints can coexist harmoniously, even if they differ in their approaches and interpretations, contributing to the cultivation of a peaceful and inclusive society.

2. Pratyaksha in Trika Shaivism

A central concept in Trika Shaivism is the emphasis on direct experiential knowledge as a path to spiritual realisation. This direct perception or immediate experience of reality is known as pratyaksha and it transcends conceptual understanding and intellectual knowledge, offering a direct encounter with the essence of existence. Trika Shaivism considers pratyaksha the highest form of knowledge because it arises from personal experience and inner realisation.

Pratyaksha is characterised by certain distinct qualities:

i. Direct and Immediate

Pratyaksha is an immediate experience of reality, unmediated by the mind or external agents. It is unfiltered by concepts, thoughts, or interpretations, providing a direct encounter with existence.

ii. Self-Verifying

Pratyaksha does not rely on external validation or confirmation. It carries its own certainty and authenticity and is inherently self-evident and self-confirming.

iii. Intuitive and Non-Conceptual

Pratyaksha transcends ordinary conceptual understanding, operating at an intuitive level. It bypasses the limitations of language and linear thinking, offering a non-conceptual awareness.

iv. Totality of Perception

Pratyaksha encompasses the totality of perception, extending beyond the confines of the sensory faculties. It includes the direct apprehension of both the outer world and the inner realms of consciousness, offering a profound awareness that transcends the limitations of the physical senses.

Trika Shaivism also acknowledges the limitations of indirect knowledge, which is acquired through inference, authority, or scriptural study. This type of knowledge relies on conceptual understanding, making it subject to interpretation, conditioning, and intellectual limitations. While indirect knowledge is valuable for understanding the teachings and practices of Trika Shaivism, it is considered secondary to the direct experiential knowledge of pratyaksha which transcends conceptual frameworks, allowing for direct communion with the true nature of reality.

Trika Shaivism provides various practices to cultivate and deepen the experience of pratyaksha. Meditation is a fundamental practice in Trika Shaivism to quiet the mind, cultivate inner stillness, and open oneself to direct experiential knowledge. Through meditation, one can enter into direct communion with the divine and access deeper levels of awareness. Another key practice is self-inquiry which involves turning inward and questioning the nature of one's own self. By exploring the question 'Who am I?' and investigating the nature of one's thoughts, emotions, and perceptions, one can directly encounter the true essence of one's being. Yogic practices, such as breath control (pranayama), energy channelling (Kundalini yoga), and body postures (asanas), are also employed to purify the body-mind complex, harmonise the energy centres (chakras), and create an optimal state for the direct experience of higher realities. Moreover, the guidance of an enlightened teacher or guru is highly valued in Trika Shaivism. A guru can transmit direct experiential knowledge through their presence, teachings, and initiations, guiding the seeker towards pratyaksha and spiritual realisation.

Pratyaksha holds transformative power in the spiritual journey of self-realisation. It brings about a radical shift in perception, dissolving the illusory boundaries between the self and the world. Through pratyaksha, one directly experiences the underlying unity and interconnectedness of all existence. It liberates the seeker from the constraints of the conditioned mind, unveiling the inherent divinity within. Furthermore, pratyaksha brings about a profound shift in understanding, challenging the limitations of ordinary knowledge. It reveals the ultimate truth of the non-dual nature of reality, where subject and object, self and other, merge into a seamless whole. This experiential realisation brings about a profound sense of liberation, freedom, and bliss.

Trika Shaivism's emphasis on direct experiential knowledge as a means of spiritual realisation offers a transformative path for seekers, transcending the limitations of conceptual understanding and providing a direct encounter with the essence of reality.

Beyond Classical Physics: Epistemological Insights from Trika Shaivism

This section delves into two pivotal aspects of quantum mechanics that significantly impact our understanding of reality: non-absolutism and the evolution of objectivity. Together, these concepts invite a re-evaluation of classical objectivity and emphasise the need for a more nuanced understanding of the quantum world.

1. **Non-absolutism in Quantum Mechanics**

Non-absolutism is a fundamental concept in quantum mechanics, reflecting the inherent uncertainty and probabilistic nature of the quantum world. This principle stems notably from Heisenberg's uncertainty principle, which asserts that certain pairs of physical properties, such as position and momentum, cannot be precisely known simultaneously. The principle implies that the more accurately one property is measured, the less accurately the other can be determined.

In quantum mechanics, non-absolutism also prompts epistemological inquiries into the nature of scientific knowledge and our capacity for precise predictions. It underscores the intrinsic limitations of our understanding due to the probabilistic nature of measurements and the uncertainty principle and challenges traditional views of determinism and objectivity It encourages a shift from seeking absolute certainty to working with probabilities and statistical predictions, embracing uncertainty and probability as fundamental aspects of quantum phenomena. This recognition emphasises the continual need for refining our scientific models and methodologies to enhance accuracy. All in all, by embracing non-absolutism, we can deepen our comprehension of the intricate nature of the quantum realm.

2. The Evolution of Objectivity

One of the most striking departures from classical physics is quantum mechanics' challenge to traditional ideas of objectivity. In classical physics, reality was viewed as independent of observation, with well-defined properties that an observer merely recorded. Quantum mechanics, however, introduces a paradigm where the relationship between observer and observed is more nuanced and intertwined with epistemological considerations. It posits measurement as an active process that influences the observed reality. This is supported by the Copenhagen interpretation which proposes that conscious observation is necessary for the collapse of the wave function, implying that consciousness actively participates in shaping observed reality. This observer-dependent measurement challenges the classical notion of an objective reality with fixed properties, injecting subjectivity into quantum phenomena.

Quantum mechanics also introduces phenomena like superposition and wave-particle duality, further challenging classical objectivity. Entanglement, another quantum phenomenon, further indicates a deeper interconnectedness that surpasses classical notions of separate, independent entities.

Quantum mechanics has spurred various interpretations that explore these challenging implications from different angles. These interpretations underscore ongoing debates about the nature of reality, the observer's role, and the interface between quantum and classical worlds. They highlight quantum mechanics' profound challenge to classical objectivity and advocate for a holistic understanding encompassing both observer and observed. Embracing these challenges invites us to expand our philosophical and scientific frameworks, striving for a deeper understanding of the fundamental nature of existence.

Epistemological Connections between Trika Shaivism and Quantum Mechanics

Trika Shaivism and quantum mechanics provide distinct yet complementary insights into the nature of knowledge, the boundaries of human understanding, and the dynamic relationship between observer and observed. This section delves into the epistemological parallels between these two domains, emphasising their mutual emphasis on direct experiential knowledge, the challenges posed to conceptual understanding, and their recognition of the interconnectedness inherent in all existence.

1. **Direct Experiential Knowledge**

Both Trika Shaivism and quantum mechanics place a strong emphasis on the importance of direct experiential knowledge in comprehending reality. In Trika Shaivism, this direct experiential knowledge is esteemed as the highest form of knowing. It surpasses mere conceptual and intellectual understanding, providing a direct encounter with the essence of existence itself. Similarly, quantum mechanics challenges the traditional view that knowledge can be solely derived from objective external observations and measurements. The process of observation and measurement in quantum mechanics suggests that our subjective interaction with the world influences our understanding of its fundamental nature. Thus, we can see that both Trika Shaivism and quantum mechanics acknowledge the inherent subjectivity in the acquisition of knowledge and underscore the profound role of direct experience in deepening our insights into reality.

2. **Interconnectedness of Observer and Observed**

Both Trika Shaivism and quantum mechanics highlight the interconnectedness and interdependence of the observer and the observed. In Trika Shaivism, the observer is viewed as an essential participant in the cosmic unfolding, where observed reality is perceived as a manifestation of consciousness itself. Similarly, quantum mechanics challenges the notion of a detached observer, suggesting that the act of observation is inseparable from the system being observed. This concept implies a profound connection between observer and observed, where the observer's interaction directly influences the observed reality, particularly evident in the process of measurement.

3. Limitations of Theoretical Frameworks

Both Trika Shaivism and quantum mechanics recognise the inherent limitations of theoretical frameworks, rationality, and language in fully grasping the nature of reality. Trika Shaivism posits that ultimate reality, represented as Shiva, transcends conceptual understanding. It teaches that language and intellectual concepts can only offer partial descriptions and indications towards the ineffable essence of reality. Similarly, quantum mechanics, relying on mathematical formalisms, acknowledges the constraints of conceptual models in comprehensively describing the complexities of quantum phenomena. Both domains encourage us to surpass the boundaries of language and rational thought to access a deeper, more profound understanding of reality.

4. Epistemological Humility

The realms of Trika Shaivism and quantum mechanics advocate for epistemological humility, recognising the inherent limitations of human knowledge. Trika Shaivism underscores the importance of transcending the intellect and ego, acknowledging that our conditioned minds impose filters and biases on how we perceive reality. Quantum mechanics, through its principle of uncertainty and the recognition of inherent measurement limits, highlights the intrinsic uncertainty and indeterminacy of the quantum realm. Together, these perspectives encourage a humble approach, reminding us that our understanding is constrained and that reality may surpass our ability to fully grasp it.

5. Contemplative Inquiry and Intuitive Insight

Trika Shaivism and quantum mechanics both emphasise the role of contemplative inquiry and intuitive insight in deepening our understanding of reality. Trika Shaivism promotes meditative practices, self-inquiry, and direct experiential realisation as pathways to profound insights. Similarly, quantum mechanics, with its abstract mathematical framework, also relies on intuition and creative thinking to comprehend its principles and achieve breakthroughs in understanding. In both disciplines, there is a shared emphasis on nurturing a contemplative mindset that fosters intuitive insights and direct engagement with the subject matter.

The epistemological parallels between Trika Shaivism and quantum mechanics provide profound insights into the nature of knowledge, perception, and reality. Both disciplines underscore the subjective nature of understanding, the constraints of rationality and language, the non-dualistic essence of reality, and the significance of epistemological humility. They also highlight the importance of contemplative inquiry, intuitive insight, and the integration of diverse perspectives. Exploring these parallels deepens our comprehension of the intricate interplay between consciousness, perception, and the pursuit of knowledge.

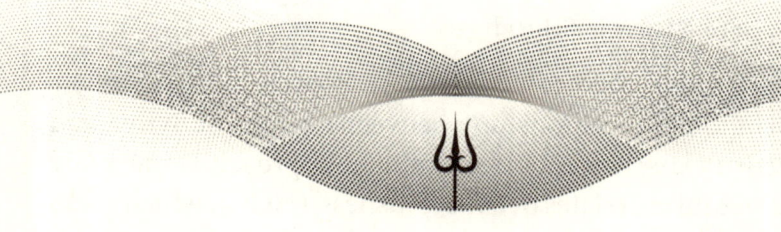

Chapter 17
Unveiling Consciousness

The nature of consciousness has been the central point of conceptual deliberation vis-à-vis the formulations of metaphysics, philosophy and science. In Indic thought, the primacy of consciousness has been highlighted in myriad traditions and treatises. We cannot bring around our discussion on the resonances of Trika Shaivism and quantum mechanics without deliberating on the import and essence of the primary element in the Trika Shaivite universe: consciousness. Consciousness, often considered the essence of subjective experience and self-awareness, has captivated human curiosity for centuries. In recent years, advancements in neuroscience, psychology, and philosophy have brought about renewed interest and deeper exploration into the nature of consciousness. This chapter will explore these perspectives in order to gain insights into the evolving understanding of consciousness and its implications for our perception of reality and our sense of self.

From Neural Networks to Integrated Information: Diverse Approaches to Understanding Consciousness

The field of neuroscience has made significant strides in unravelling the neural correlates of consciousness and exploring the relationship between brain activity and subjective experience. One prominent approach is the study of the 'neural correlates of consciousness' (NCC), which seeks to identify brain processes that are directly associated with conscious awareness. Functional neuroimaging techniques, such as fMRI and EEG, have shed light on the neural networks involved in various conscious states, including wakefulness, sleep, and altered states of consciousness. Neuroscientists also investigate the role of specific brain regions, such as the prefrontal cortex and the default mode network, in supporting self-referential thought and introspection.

While neuroscientific approaches provide valuable insights into the neural underpinnings of consciousness, they still face significant challenges in bridging the hard problem of consciousness—explaining why and how subjective experience arises from objective brain processes. This has spurred debates on whether a purely reductionist approach can fully account for the richness and qualia of conscious experience.

In parallel, we have now reached the age of artificial and quantum neural networks. Artificial neural networks (ANNs) and quantum neural networks (QNNs) are two distinct approaches to machine learning that hold great promise for solving complex problems and advancing our understanding of artificial intelligence. While ANNs have been widely adopted and proven successful in various applications, QNNs harness the principles of quantum mechanics to offer unique advantages and capabilities. Let's look at how these neural networks can provide valuable insights into the nature of consciousness.

ANNs are computational models inspired by the structure and functioning of the human brain. They consist of interconnected artificial neurons or nodes organised into layers. Each node receives inputs, applies a mathematical function to them, and produces an output. Through a process called training, ANNs learn to recognise patterns and make predictions based on a large amount of labelled data. QNNs, on the other hand, are a novel class of machine learning models that leverage the principles of quantum mechanics to perform computations. QNNs use quantum bits or qubits, which can exist in superposition states, to encode and process information. Qubits allow for the simultaneous exploration of multiple possibilities, providing significant computational advantages over classical binary bits. QNNs have the potential to solve complex problems more efficiently than classical ANNs in certain domains. They can exploit quantum entanglement, where the state of one qubit is linked to the state of another, to perform computations in parallel and tackle combinatorial optimisation problems more effectively. QNNs also offer the possibility of enhanced data representation and increased resilience to noise.

The relevance of such artificial constructs is even more fundamental when one looks at conceptualisations of consciousness in terms of modern frameworks like Tononi's integrated information theory. Integrated information theory (IIT), proposed by neuroscientist Giulio Tononi, is a comprehensive theoretical framework that seeks to explain the nature of consciousness. It offers a unique perspective on how consciousness emerges from the integration of information within complex systems. At the heart of Integrated information theory is the concept of integrated information, denoted by the symbol Φ (phi). According to Tononi, a system possesses consciousness to the extent that it contains a large amount of integrated information. Integrated information is not just the sum of its parts but arises from the interplay and integration of its components. Tononi argues that consciousness is not limited to humans or animals but can potentially exist in any complex system that exhibits a high degree of integrated information. This perspective challenges anthropocentric views of consciousness and opens up the possibility of ascribing consciousness to other entities, such as artificial intelligence systems or even certain physical systems.

One of the key principles of integrated information theory is that consciousness is irreducible and cannot be explained solely by analysing its constituent parts. Tononi posits that consciousness is a fundamental property of complex systems and cannot be reduced to the behaviour of individual components or the mere presence of information. Instead, consciousness arises from the way information is integrated within the system. Another central concept in integrated information theory is the notion of 'maximally irreducible cause-effect structure'. Tononi argues that conscious experiences are characterised by a high level of irreducibility, meaning they cannot be broken down into independent parts without losing their specific qualities. This irreducibility distinguishes conscious experiences from mere aggregates of information. Furthermore, integrated information theory states that the quality and richness of conscious experience are determined by the degree of integrated information within a system. Systems with high Φ values exhibit more complex and differentiated conscious states, while systems with low Φ values have simpler and less differentiated experiences. This aspect of the theory provides a measure of consciousness, allowing for quantitative assessments of the level of conscious awareness within different systems.

While IIT offers a novel and comprehensive framework for understanding consciousness, it is not without its criticisms. Some researchers argue that the theory relies heavily on subjective assessments and lacks empirical validation. Others suggest that the theory does not fully account for the dynamic and distributed nature of information processing in the brain. Nevertheless, Tononi's theory has sparked significant interest and debate within the field of consciousness studies. It provides a comprehensive framework for understanding the relationship between information integration and conscious experience, offering new avenues for further exploration and research into the enigmatic nature of consciousness.

Philosophy has long been another major field dedicated to exploring and addressing questions about consciousness. Different philosophical schools offer diverse perspectives on the nature and metaphysical implications of consciousness. For instance, dualism posits that consciousness is fundamentally distinct from the physical body, while monism suggests that consciousness and physical phenomena are different aspects of a unified whole. Idealism proposes that consciousness is primary, and the material world arises from it, while panpsychism suggests that consciousness is inherent to all matter. Philosophical explorations of consciousness also delve into subjective aspects, such as qualia (the qualities of subjective experience), intentionality (the directedness of consciousness toward objects), and the problem of why and how consciousness arises. These inquiries contribute to the broader understanding of consciousness and its relationship to the nature of reality.

To address the complexity of consciousness, many modern approaches seek to integrate scientific and philosophical insights, recognising the need for interdisciplinary exploration. Cognitive science, for example, draws on psychology, neuroscience, and computer science to study the computational and cognitive aspects of consciousness. The study of 'embodied cognition' emphasises the role of the body, sensorimotor processes, and environmental interactions in shaping conscious experience. Another interdisciplinary field, known as 'consciousness studies', investigates the nature of consciousness through multiple lenses, including neuroscience, psychology, philosophy, and contemplative practices. This integrative approach acknowledges the limitations of reductionism and aims to bridge different disciplines to gain a more comprehensive understanding of consciousness.

Moreover, many non-dualistic perspectives on consciousness draw inspiration from both Eastern and Western philosophies, challenging the notion of a separate self and emphasising the interconnectedness of all phenomena. Advaita Vedanta, a Hindu philosophy, proposes that ultimate reality is non-dual, and individual consciousness is an expression of the universal consciousness. Similarly, some contemporary thinkers draw upon insights from Buddhist philosophy, highlighting the interdependence and impermanence of all things. These non-dualistic perspectives suggest that consciousness is not confined to an individual entity but is an intrinsic aspect of the fabric of reality. They offer alternative frameworks for understanding the nature of consciousness and its relationship to the world, emphasising the interconnectedness of all existence.

Quantum physics, with its inherent indeterminism and non-locality, has been employed in probing the nature of consciousness. Some theories propose that consciousness may play a role in the collapse of the quantum wave function, influencing the measurement outcomes. Speaking on the quantum measurement problem, Freeman Dyson has stated:

> 'Unfortunately, people writing about quantum mechanics often use the phrase "collapse of the wave function" to describe what happens when an object is observed. This phrase gives a misleading idea that the wave function itself is a physical object. A physical object can collapse when it bumps into an obstacle. But a wave function cannot be a physical object. A wave function is a description of a probability, and a probability is a statement of ignorance. Ignorance is not a physical object, and neither is a wave function. When new knowledge displaces ignorance, the wave function does not collapse; it merely becomes irrelevant.'

Recent research has explored the possibility of even stronger nonlocal correlations beyond those allowed by quantum mechanics. This raises questions about whether there are fundamental physical principles that limit the strength of nonlocality in nature. However, the relationship between consciousness and quantum mechanics remains a topic of debate and speculation, and further research is needed to establish concrete connections.

In summary, modern perspectives on consciousness encompass a wide range of scientific, philosophical, interdisciplinary, and non-dualistic approaches. The study of consciousness has advanced our understanding of the neural mechanisms associated with subjective experience, deepened philosophical inquiries into the nature of consciousness, and fostered interdisciplinary collaborations. While many questions remain, the exploration of consciousness is a dynamic and evolving field, inviting us to reevaluate our assumptions about the nature of reality, the self, and our place in the universe. As scientific and philosophical investigations continue, the quest to unravel the mysteries of consciousness promises to shed light on fundamental aspects of our existence.

Trika Shaivism's Perspective on Consciousness

Trika Shaivism offers a profound perspective on the nature of consciousness and ultimate reality. According to this philosophical and spiritual tradition, consciousness is dynamic and constantly in flux, rather than static or fixed. It recognises that within the ultimate reality, consciousness manifests through various movements, vibrations, and fluxes. These expressions of consciousness are inherent aspects of its nature, not separate from it. Trika Shaivism also encourages individuals to attune themselves to the dynamic nature of consciousness and embrace the ever-changing flow of experience.

According to Trika Shaivism, there are three primary fluxes or movements of consciousness within the ultimate reality: the flux of knowing, the flux of willing, and the flux of feeling. These fluxes represent different aspects of consciousness and their interplay in shaping our experience of reality. The flux of knowing involves perception, cognition, and awareness. The flux of willing encompasses intention, desire, and action. The flux of feeling encompasses emotions, sensations, and subjective experiences. Trika Shaivism recognises that these fluxes are interconnected and inseparable, constituting the tapestry of human experience.

Trika Shaivism's understanding of consciousness fluxes is intricately linked with the concept of Shiva and Shakti—

divine principles representing consciousness and energy, respectively. It portrays their interplay as a cosmic dance, where consciousness fluxes emerge from the dynamic interaction of these cosmic energies. Shiva symbolises pure consciousness, while Shakti embodies the creative energy that manifests and sustains the universe. This dance of Shiva and Shakti gives rise to the fluxes of consciousness and the diverse experiences within ultimate reality. Trika Shaivism emphasises the importance of recognising and aligning with these fluxes of consciousness to deepen one's understanding of ultimate reality. By acknowledging the dynamic nature of consciousness and its various fluxes, individuals can cultivate a heightened awareness of their own experience.

Trika Shaivism encourages individuals to observe the fluxes of knowing, willing, and feeling as they arise within awareness, fostering mindfulness and a keen perception of the ever-changing flow of consciousness in each moment. While acknowledging these fluxes as integral to human experience, Trika Shaivism also stresses the importance of transcending attachment to them. The ultimate goal is to realise the essential nature of consciousness beyond these fluxes and to abide by the unchanging awareness underlying all experiences. Trika Shaivism teaches that while fluxes of consciousness may arise and fade, pure consciousness—the ultimate reality—remains constant. By transcending identification with these fluxes, individuals can attain deeper peace, freedom, and liberation.

Trika Shaivism offers various practices aimed at developing awareness and facilitating transformation concerning the fluxes of consciousness. These practices encompass meditation, self-inquiry, mantra repetition, and devotional practices. By engaging in these practices, individuals can heighten their awareness of the fluxes of consciousness, observe how these fluxes influence thoughts, emotions, and actions, and consciously align themselves with the divine flow. They enable a profound understanding of consciousness's dynamic nature and support the journey toward self-realisation and union with the ultimate reality.

Overall, Trika Shaivism offers a profound understanding of the dynamic nature of existence through its perspective on the fluxes of consciousness within the ultimate reality. By acknowledging the ever-changing flow of consciousness and its diverse manifestations, individuals can deepen their awareness of personal experience and harmonise with the divine dance of Shiva and Shakti. Through practices focused on awareness and transcendence, individuals can move beyond identification with these fluxes and realise the unchanging essence of consciousness that underpins all experiences.

Drawing Parallels: Trika Shaivism and Quantum Mechanics' Concept of Consciousness

In this section, we will explore the shared insights into the nature of consciousness, when it comes to Trika Shaivism and quantum mechanics. By examining these resonances, we can gain a deeper understanding of the potential bridges between ancient spiritual wisdom and modern scientific inquiry.

1. Exploration of Consciousness

Trika Shaivism offers detailed insights into various states of consciousness, providing methodologies such as meditation, mantra recitation, and contemplative practices to transcend ordinary awareness and access higher realms. Similarly, certain interpretative traditions in quantum physics encourage us to broaden our understanding of consciousness and perception through phenomena like wave-particle duality and superposition. Both suggest a shared interest in exploring what it means to be 'conscious', either with the human mind or even with non-anthropomorphised pointer states!

2. Consciousness and Unity

Trika Shaivism recognises the fundamental oneness of consciousness, where all beings are interconnected and inseparable from the divine source. Quantum mechanics, through phenomena like entanglement, demonstrates particles can be intricately linked regardless of distance, highlighting the underlying unity of the quantum realm. This convergence invites recognition of interconnectedness among all individuals and societies.

3. Self-Transformation and Consciousness

Trika Shaivism emphasises exploring consciousness for self-transformation, recognising true change comes from within as individuals cultivate awareness. Some interpretations of quantum mechanics suggest that consciousness plays a fundamental role in shaping reality through observation and wave function collapse. This highlights the power of correlations, 'conscious' participation and agency in effecting individuated and group transformation.

4. Non-Dualistic Consciousness

Trika Shaivism emphasises the non-dualistic nature of reality, where apparent opposites are expressions of the same underlying consciousness. Quantum mechanics challenges binary thinking through wave-particle duality and superposition. This convergence invites adopting a non-dualistic perspective in understanding consciousness and reality.

5. Consciousness and Ethics

Trika Shaivism emphasises recognising the inherent divinity and worth of all beings through consciousness. Quantum mechanics highlights the potential impact of our conscious actions on the broader web of existence through entanglement. This convergence calls for re-evaluating ethical frameworks based on an expanded understanding of consciousness and its role in reality.

The connections between Trika Shaivism and quantum physics offer a rich and fertile ground for further exploration and inquiry. These shared insights, despite originating from vastly different traditions, offer a rich ground for contemplation on the nature of reality, consciousness, and our place in the universe. They invite us to expand our perspective beyond conventional boundaries, fostering a more holistic and interconnected understanding of existence.

Acknowledgements

The meditations comprising *From Shiva to Schrödinger* were made possible through the steadfast support and encouragement of many individuals. I extend my sincere gratitude to my mentors and teachers, whose guidance has been invaluable.

Prof. Brian Josephson, Nobel laureate in Physics (1973) and one of the 20th century's most distinguished physicists, has been instrumental in my journey. Despite his renown, he remains remarkably humble. Our discussions went beyond the complexities of non-linear dynamics and condensed matter physics to explore mind-matter unification and the fundamental nature of reality. In a universe increasingly viewed through the lens of information dynamics—where principles and meta-dynamics might be more fundamental than the traditional interplay of mass and energy—we investigated coordination dynamics and the importance of correlations in nature.

I also wish to acknowledge Prof. Crispin Barnes, Prof. Roger Penrose, Prof. Dipankar Home, Prof. Sisir Roy, Prof. Prasanta Panigrahi, Prof. Subhash Kak, Prof. Scott Kelso, Prof. S. V. Eswaran, Prof. Ravindra Pratap Singh, Prof. Konepudy Sreenivas, and Prof. Anand Ranganathan for the privilege of discussing the foundations of contemporary scientific frameworks, particularly quantum physics.

I would like to pay my respects to Sri Ramakrishna Paramhansa, whose grace, conveyed through Swami Sarvasthananda Maharaj, my spiritual preceptor, has guided me in exploring what I perceive as a vast realm of wisdom and realisation. Thakur seamlessly integrated jñāna yoga and Vedanta with bhakti and tantra. I also wish to acknowledge Swami Sarvapriyananda Maharaj, Swami Atmapriyananda Maharaj, and Ayon Maharaj for their contributions.

I would like to honour and acknowledge the legacy of Shri Vasugupta, Shri Ādi Śaṅkarācārya, Shri Utpaladeva, Shri Abhinavgupta, Shri Rajanaka Kṣemarāja, Shri Somānanda, Swami Lakshman Joo, and Acharya Rameshwar Jha for their inspiration in this work.

I extend my gratitude to Shri Rajiv Malhotra for guiding me in developing intellectual Kshatriyata—standing with courage and conviction for our civilisational ethos and wisdom in our quest for decolonisation. My thanks also go to Prof. K. Ramasubramanian for his invaluable insights into Nyāya-Vaiśeṣika and Vedanta, and to Prof. R. Vaidyanathan, Prof. Kantilal Mardia, Prof. Rama Jayasundar, Pandit Satish Sharma, and Shri Jay Lakhani for our discussions on various aspects of Indian knowledge systems.

I would like to acknowledge Shri Jayant Sahasrabudhe for our discussions on non-absolutism in physicalist heuristics, uncertainty, and correlations. His support was instrumental in establishing Mandala, an international platform dedicated to exploring ontological and epistemological resonances between modern science and Bharatiya Jñāna Paramparā, under the aegis of Vijñāna Bharati.

I am also grateful to Shri Chamu Krishna Shastry for emphasising the importance of acquiring linguistic tools for a meaningful exploration of Bharatiya Jñāna Paramparā—a pursuit that continues, given the vast expanse of Sanskrit.

My heartfelt thanks go to my wonderful family: Ma—Dr. Karabi M. G. Majumdar, Baba—Prof. Rupendra Guha Majumdar, my wife Devyani Guha Majumdar, and my brother Tirthankar Guha Majumdar. Their unwavering support and patience throughout the creation of this book have been invaluable. Their belief in my vision provided the strength and motivation needed to complete this work.

I would also like to honour my forbears, particularly Aita, Thakurma, Dadu, and Koka Deuta, whose *sāṃskāra* have shaped who I am today. My gratitude extends to my avunculus, Prof. Arup Kumar Misra, for his significant contributions to my journey.

I offer my heartfelt thanks to my publishers, Hay House India, for their belief in this project and their unwavering support in bringing this book to life. Their professionalism and commitment have ensured that this work reaches a wider audience.

Lastly, I extend sincere appreciation to all readers and seekers who engage with this book. I hope it sparks curiosity and fosters a deeper understanding of the profound connections between Trika Shaivism and quantum physics.

Om Namah Shivaya.

With gratitude,
—Dr Mrittunjoy Guha Majumdar

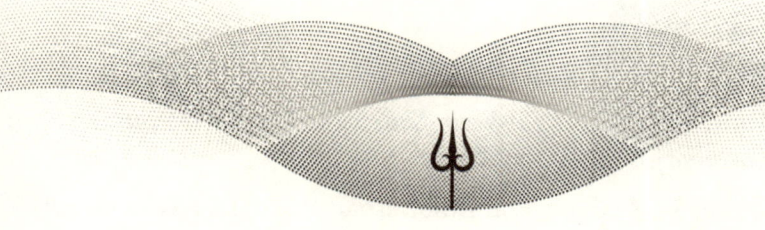

References

Chatterji, Jagadish Chandra. *Kashmir Shaivaism*. State University of New York Press, 1962.

Shankarananda, Swami. *The Yoga of Kashmir Shaivism: Consciousness Is Everything*. Motilal Banarsidass, 2016.

Pandit, B. N. *Specific Principles of Kashmir Shaivism* (3rd ed., 2008), p. 79.

Singh, Jaideva. *Spanda-Karikas: The Divine Creative Pulsation; The Karikas and the Spanda-Nirnaya*. Motilal Banarsidass, 2014.

Dyczkowski, Mark S. G. *The Doctrine of Vibration: An Analysis of the Doctrines and Practices Associated with Kashmir Shaivism*. State University of New York Press, 1987.

Dhar, A. N. 'A Reappraisal of Lal Ded.' *Himalayan and Central Asian Studies* 11.2 (2007): 49.

Bazaz, Prem Nath. 'Influence of Shaivism on Nund Rishi.' *Indian Literature* 16.1/2 (1973): 256-267.

Singh, Jaideva. *Para-trisika-Vivarana of Abhinavagupta: The Secret of Tantric Mysticism*. Motilal Banarsidass, 2014.

Kaw, Maharaj K. *Kashmir and Its People: Studies in the Evolution of Kashmiri Society*. Vol. 4. APH Publishing, 2004.

Kaul, Shonaleeka. *The Making of Early Kashmir: Landscape and Identity in the Rajatarangini*. Oxford University Press, 2018.

Jee, Swami Lakshman. *Kashmir Shaivism: The Secret Supreme.* State University of New York Press, 1988.

Toshakhānī, Śaśiśekhara, and Kulbhushan Warikoo, eds. *Cultural Heritage of Kashmiri Pandits.* Pentagon Press, 2009.

Sharma, Niharika. 'The Trika School—A Religio-Philosophical Emergence.' *Tattva: Journal of Philosophy* 13.2 (2021): 41-58.

Dhar, K. N. 'Abhinavagupta: The Philosopher.' *Saints and Sages of Kashmir* 5 (2004): 53.

Shastri, Mukund Ram. *Tantraloka*, with the Viveka Commentary of Jayaratha. 1918.

Misra, Nityananda. *The Om Mala: Meanings of the Mystic Sound.* Bloomsbury Publishing, 2018.

Gates, Sarah Louise. 'Pragmatic Non-Duality in William James, Swāmī Vivekānanda, and Trika Shaivism.' *Pragmatism, Spirituality and Society: New Pathways of Consciousness, Freedom and Solidarity* (2021): 41-64.

Arraj, William. 'The Triadic Heart of Śiva: Kaula Tantricism of Abhinavagupta in the Non-Dual Shaivism of Kashmir.' 1991: 175-178.

Panigrahy, Satyabrata. *The Land of Yoga.* Blue Hill Publications, 2021.

Ghosh, Monidipa. 'The Nexus between Music and Kaśmīra Śaivism.' *Sangeet Galaxy* 11.1 (2022).

Rambachan, Anantanand. 'Hierarchies in the Nature of God? Questioning the 'Saguna-Nirguna' Distinction in Advaita Vedanta.' *Journal of Hindu-Christian Studies* 14.1 (2001): 7.

Paglione, Johnpierre, and Richard L. Greene. 'High-Temperature Superconductivity in Iron-Based Materials.' *Nature Physics* 6.9 (2010): 645-658.

Danylova, T. V. 'Searching for the True Self: The Way of Nondual Wisdom.' *Антропологические Измерения Философских Исследований* 12 (2017): 7-15.

Chihaia, Ştefania. 'The Case for Working with Feminist New Materialisms Against the Dualisms That Divide Us.' *Journal of International Women's Studies* 25.2 (2023): 12.

Huseyinzadegan, Dilek. 'On Hegel's Radicalisation of Kantian Dualisms: The Debate between Kant and Hegel.' *Hegel-Jahrbuch* 2015.1 (2015): 149-154.

Laycock, Steven W. 'Telic Divinity and Its Atelic Ground.' *From the Sacred to the Divine: A New Phenomenological Approach*. Dordrecht: Springer Netherlands, 1994. 43-54.

Hardie, William F. R. 'Aristotle's Treatment of the Relation between the Soul and the Body.' *The Philosophical Quarterly (1950-)* 14.54 (1964): 53-72.

Barney, Rachel, Tad Brennan, and Charles Brittain, eds. *Plato and the Divided Self*. Cambridge University Press, 2012.

Ryle, Gilbert. *The Concept of Mind*. Routledge, 2009.

Nagel, Thomas. 'Brain Bisection and the Unity of Consciousness.' *Synthese* (1971): 396-413.

Husserl, Edmund. *Ideas: General Introduction to Pure Phenomenology*. Routledge, 2012.

Votsis, Athanasios. 'Michel Foucault's Moral Subjectivity and the Semiotic Modeling of Knowledge.' *Semiotica* 2012.192 (2012): 243-250.

Derrida, Jacques. 'Structure, Sign, and Play in the Discourse of the Human Sciences.' *MA English* (1970): 51.

Lloyd, Moya. *Judith Butler: From Norms to Politics*. Vol. 20. Polity, 2007.

Caputo, John D., ed. *Deconstruction in a Nutshell: A Conversation with Jacques Derrida*. Vol. 53. New York: Fordham University Press, 1997.

Hesse-Biber, Sharlene Nagy. 'Feminist Research.' *The SAGE Encyclopedia of Qualitative Research Methods* (2008): 339-340.

Guschke, Bontu Lucie. 'Solidarity Across Difference–Rethinking Transformational Critique from Black Feminist and Postcolonial Perspectives.' *Intersektionale Solidaritäten*.

Whiu, Leah. 'Resistance and Becoming: The Inevitable Paradox of Oppression.' *Yearbook of New Zealand Jurisprudence* 5 (2001): 113-149.

Baker, Gordon P., Gordon Baker, and Katherine J. Morris. *Descartes' Dualism*. Psychology Press, 2002.

Berman, David. 'George Berkeley: Idealism and the Man.' *The Scriblerian and the Kit-Cats* 28.1 (1995): 107.

Atherton, Margaret, ed. *The Empiricists: Critical Essays on Locke, Berkeley, and Hume*. Rowman & Littlefield, 1999.

Williams, Michael. 'Descartes and the Metaphysics of Doubt.' *Essays on Descartes' Meditations* (1986): 117-139.

Kant, Immanuel. *An Answer to the Question: What Is Enlightenment?* Penguin UK, 2013.

Smith, Dorothy E. 'Telling the Truth after Postmodernism 1.' *Symbolic Interaction* 19.3 (1996): 171-202.

Mlodinow, Leonard, and Stephen Hawking. *The Grand Design*. Random House, 2010.

Sharma, B. N. Krishnamurti. *A History of the Dvaita School of Vedānta and Its Literature*. Vol. 1. Motilal Banarsidass Publishers, 1960.

Rambachan, Anantanand. *The Advaita Worldview: God, World, and Humanity*. SUNY Press, 2006.

Jones, Richard H. 'Vidyā and Avidyā in the Īśa Upanishad.' *Philosophy East and West* (1981): 79-87.

Koller, John M. 'Knowledge and Reality: Nyaya-Vaisheshika.' *Oriental Philosophies*. Macmillan Education UK, 1985. 70-82.

Richards, Glyn. 'Śūnyatā: Objective Referent or Via Negativa?' *Religious Studies* 14.2 (1978): 251-260.

Wayman, Alex. 'Buddhist Dependent Origination.' *History of Religions* 10.3 (1971): 185-203.

Stroud, Scott R. 'Anekāntavāda and Engaged Rhetorical Pluralism: Developing Cross-Cultural Sensitivity with a Jain Practice.' *Rhetoric Review* 34.2 (2015): 129-145.

Gupta, Ravi. *The Bhagavata Purana: Sacred Text and Living Tradition*. Columbia University Press, 2016.

Sokolowski, Robert. 'The Status of the Verbum in Augustine and Aquinas.' *Journal of the History of Philosophy* 14.3 (1976): 315-333.

Stump, Eleonore. 'The Problem of Evil.' *Faith and Philosophy* 2.4 (1985): 392-423.

Feenberg, Andrew. 'On the 'Roots' of Loneliness.' *Contributions to the Problem of Loneliness* (1983): 44.

MacFarquhar, Larissa. *Strangers Drowning: Grappling with Impossible Idealism, Drastic Choices, and the Overpowering Urge to Help*. Penguin, 2015.

Barrow, John D., and Frank J. Tipler. *The Anthropic Cosmological Principle*. Oxford University Press, 1986.

Guha Majumdar, Mrittunjoy, and Brian Josephson. 'Unified Field: How Self-Selected Fluctuations Can Underlie Reality' (2020).

Ullah, Arif, and Pavlo O. Dral. 'MLQD: A Package for Machine Learning-Based Quantum Dissipative Dynamics.' *Computer Physics Communications* 294 (2024): 108940.

Zheng, Guo, et al. 'Near-Optimal Performance of Quantum Error Correction Codes.' *Physical Review Letters* 132.25 (2024): 250602.

Popovic, Maria, Mark T. Mitchison, and John Goold. 'Thermodynamics of Decoherence.' *Proceedings of the Royal Society A* 479.2272 (2023): 20230040.

Moore, Henrietta L., and Todd Sanders. 'Anthropology and Epistemology.' *Anthropology in Theory: Issues in Epistemology* (2006): 1-21.

Maffie, James. 'Towards an Anthropology of Epistemology.' *Philosophical Forum* 26.3 (1995).

CONNECT WITH
HAY HOUSE
ONLINE

 hayhouse.co.in   @hayhouseindia

 @hayhouseindia   @hayhouseindia

Join the conversation about latest products, events, exclusive offers, contests, giveaways and more.

'*The gateways to wisdom and knowledge are always open.*'

Louise Hay

www.ingramcontent.com/pod-product-compliance
Lightning Source LLC
LaVergne TN
LVHW041659070526
838199LV00045B/1121